PROFESSIONAL
NFC APPLICATION DEVELOPMENT
FOR ANDROID™

PROFESSIONAL

NFC Application Development for Android™

PROFESSIONAL

NFC Application Development
for Android™

Vedat Coskun
Kerem Ok
Busra Ozdenizci

wrox™
A Wiley Brand

My beloved love; Istanbul, the magnificent
I am so lucky to be born out of you,
and my passionate ambition is to be buried into you
as well.

—Vedat Coskun

To my dear family.

Her zaman yanımda olan aileme.

—Kerem Ok

To my lovely family and Ugurcan who encouraged me
to do my best.

—Busra Ozdenizci

ABOUT THE AUTHORS

 **VEDAT COSKUN** is a computer scientist, academician, and author. He established NFC Lab – İstanbul (www.NFCLab.com), the leading research lab on Near Field Communication (NFC) technology worldwide, which aims to take initiative on sustainable evolution of the technology for creating a win-win ecosystem for all the actors in the game such as users and financial and technical organizations. He is currently working as Associate Professor of Information Technology at ISIK University, Istanbul. He received the "Excellence in Teaching" award from ISIK University in 2012. He also gave lectures at several other universities such as University of Thessaly in Volos, Greece; Malardalen University in Vasteras, Sweden, and Inholland University in Amsterdam, Netherlands. He specializes in security, mobile technologies, Java technology, Android, and NFC. He has written a vast amount of conference and journal publications, and authored several books, including *Near Field Communication (NFC): From Theory to Practice* (Wiley, 2012). He believes that establishing a strong relationship between academia and the NFC industry is important, and considers his role as a consultant for national and international companies as a catalyst to making that happen.

 **KEREM OK** is a PhD candidate in the Informatics department at Istanbul University. His research areas are NFC, mobile technologies, web technologies, and mobile usability. He has authored several journal and conference publications on NFC technology. He is also one of the authors of *Near Field Communication (NFC): From Theory to Practice*. He is currently a researcher at NFC Lab – İstanbul.

 **BUSRA OZDENIZCI** received her MS degree in Information Technologies from ISIK University, Turkey, and is pursuing her PhD degree in the Informatics department at Istanbul University. Her research areas include NFC, mobile communication technologies, and mobile persuasion. She has authored several conference and journal publications on NFC technology. She is one of the authors of the book titled *Near Field Communication (NFC): From Theory to Practice*. She is currently a researcher at NFC Lab – İstanbul.

ABOUT THE TECHNICAL EDITORS

PETR MAZENEC is the cofounder of the Mautilus, s.r.o. company, which is focused on NFC technology and custom software development for smartphones and tablets. He is currently responsible for coordinating NFC activities and technical project leadership for Smart TV development. Petr became interested in computers in the late 80s, when he was one of the few lucky users of the Commodore 64 machine behind the iron curtain. He started programming at that time and since then software development has become his passion. He has progressed from coding on the assembler in MS-DOS up to the current development of software for the most recent smartphone platforms. He started mobile development in 2003 on the Symbian platform, when Nokia released the first smartphone 7650 followed by the famous Siemens SX1. As a Symbian developer, Petr participated in and won several developers' competitions and was named a Forum Nokia Champion six times in a row.

HANK CHAVERS is Associate Principle at Constratus, a consultancy providing technical expertise and business analysis for telecommunications, where he is leading the NFC innovation efforts with key clients. Hank has over 20 years of experience in development, deployment, and converging of Internet services and wireless data. He has advised and consulted many companies — including CNN, ESPN, and Sabre — in expanding their products to mobile; and AT&T, T-Mobile, and Verizon Wireless in launching rich data services. His NFC-specific accomplishments include: producing proof-of-concept demonstrations for NFC-enabling wireless technologies; providing technical management for NFC trials, including the first trial conducted with two types of payment cards loaded on one device; and leading the NFC Forum Developer Workgroup and NFC Global Competition.

CREDITS

**VP CONSUMER AND TECHNOLOGY
PUBLISHING DIRECTOR**
Michelle Leete

**ASSOCIATE DIRECTOR—BOOK CONTENT
MANAGEMENT**
Martin Tribe

ASSOCIATE PUBLISHER
Chris Webb

ASSOCIATE COMMISSIONING EDITOR
Ellie Scott

ASSOCIATE MARKETING DIRECTOR
Louise Breinholt

MARKETING MANAGER
Lorna Mein

SENIOR MARKETING EXECUTIVE
Kate Parrett

EDITORIAL MANAGER
Jodi Jensen

SENIOR PROJECT EDITOR
Sara Shlaer

PROJECT EDITOR
Kathryn Duggan

EDITORIAL ASSISTANT
Annie Sullivan

TECHNICAL EDITORS
Petr Mazenec
Hank Chavers

PRODUCTION EDITOR
Christine Mugnolo

COPY EDITORS
Chuck Hutchinson
Grace Fairley

PROOFREADER
Sarah Kaikini

INDEXER
Robert Swanson

COVER DESIGNER
LeAndra Young

COVER IMAGE
© fatih donmez / iStockphoto

CONTENTS

INTRODUCTION

MOBILE PHONE TECHNOLOGY has been in a race in recent years to integrate new technologies and services, and the actors involved are all striving to be in the leading group that proposes new suggestions to the users. Innovative additional services entice users, who try to beat, or at least catch up with the people around them. Young people are especially keen to be part of such competition. Adults, on the other hand, aim to use the most efficient services to make their lives easier — and to be a little bit admired at the same time.

In terms of the appetite for using new technologies, companies do not lag behind the users. They are aware that companies that take the lead in promoting new technologies by embedding them in new services and offering them to the users will come out ahead, and that this is extremely important in today's competitive world. Most companies try to propose new services themselves, if possible, or by a minimal number of companies working together if it is not. They try to entice the user by offering them services with low costs, and enhanced with additional features.

Until recently, Near Field Communication (NFC) was not known at all. In just in a few years it has been introduced with great enthusiasm by organizations including governmental departments, research centers, and companies.

There are two major areas in which NFC has the potential for success. The first is its technological sufficiency; the other is the ecosystem agreement by the actors in the game. These are very much interrelated. As the actors become convinced about the success of the new model, they invest more resources to develop it; and as new technical improvements take place, the ecosystem becomes more established and ready for the boom. When one actor invests more money in this option, that actor becomes more eager to make agreements with other actors in order to recoup their funding and achieve a better return on investment (ROI). When all the factors are analyzed, it might be confidently suggested that an NFC boom is now about to start.

As a short-range wireless communication technology that potentially facilitates the mobile phone usage of billions of people over the world, NFC offers an enormous number of use cases — including credit cards, debit cards, loyalty cards, car keys, and access keys to hotels, offices, and houses — and has the potential eventually to integrate all such materials into one single mobile phone. NFC is already having an enormous impact on the financial ecosystem, as well as on mobile technology throughout the world. Mobile phone manufacturers, mobile network operators (MNOs), financial institutions such as banks, and information technology firms are performing R&D activities to increase their share of the pie as much as possible.

NFC has become a real innovation in today's mobile technology. Despite the fact that the technical structure of NFC is so simple, it offers a huge array of services, which is very important when you consider the ecosystem point of view. Potentially, it promises a vast number of ways to reach mobile phone users. Payment seems the foremost option for attempting to internalize NFC technology to the portfolio of promising services. Loyalty is another attractive way to entice users, since traditional loyalty services are already so common. Social media looks like the next promising area in which to expose new services, considering the huge explosion in social media use in recent years.

When users purchase an NFC-enabled mobile phone, they are curious about how to make use of the new annex to the traditional phone, and immediately try to do so. Many try to learn how to use NFC capability by touching their phone to another NFC-enabled phone, or other wireless technology devices. They are not aware that a program enabling a particular service has to be installed on the phone for this purpose. This is one of the shortcomings of NFC technology. When a service is embedded into the mobile phone, such as a movie camera, the user catches up very quickly if he or she is already acquainted with movie recorders. NFC, on the other hand promises new services that the ordinary user is not familiar with. Hence, some form of training will be required.

NFC technology is marvelous in the sense that almost everybody can design, at least amateurishly, many new services. Some NFC-enabled mobile phones offer development services to ordinary users, mostly to make money. There is no problem with this, because NFC presents a convenient opportunity for potential entrepreneurs. One very important point here is the need to be aware that many services require collaboration with companies — sometimes large companies — which might not be eager to invest in people who try to muscle in. The payment sector, in particular, requires the co-operation of huge companies such as banks, and hence is not suitable for individual entrepreneurs.

This book will give the reader a solid and complete understanding of NFC technology, NFC application development essentials on Android technology, and NFC business ecosystem. We provide information on NFC technology (i.e., NFC operating modes and technical essentials), an introduction to Android programming technology, NFC programming essentials on Android technology, short use cases and case studies, application development phases, and NFC business ecosystem and business model alternatives with some examples over the world. With this book, solid information on NFC technology and application development is provided that meets the needs of people who are interested in NFC technology and its ecosystem, or practitioners interested in developing NFC projects.

NFC LAB – İSTANBUL

NFC Lab – İstanbul (www.NFCLab.com) considers NFC an emerging technology that transforms innovative ideas into reality for the information and communication society of the future.

This book is the collective effort of the researchers of NFC Lab – İstanbul. We as the researchers of NFC Lab- İstanbul are committed to working on NFC technology with a multidisciplinary network of expertise all around the world.

NFC Lab – İstanbul strives for research excellence in focused research areas relevant to NFC. The lab is aimed to be a catalyst in achieving substantial progress with the involvement of key players including mobile network operators (MNOs), financial institutions, government agencies, other research institutes, trusted third parties, and other service providers. The core team is accountable for creating and maintaining the business and academic partnerships and dynamically generates networks on a project basis.

WHO THIS BOOK IS FOR

When a practitioner with some expertise in programming in Java decides to access this new area, the most they can do is try to find the required information on Java from different sources and then try to merge it. This will not be simple, because in order to build NFC applications using Java language, the practitioner needs to collect scattered information, and then merge it for a better understanding. Even in this case, the amount of information the user would collect would be very small indeed. Some basic information exists in the public domain, but much more exists only in academic literature, which is either not publicly available or not easy for non-academic people to combine with the public information. Although some basic information exists in the current literature, there is much information that is not yet available at all. For example, we have performed extensive ecosystem analysis in this work and hence recognize the lack of and need for a solid source that contains accurate information and addresses entrepreneurs and programmers.

This book is for anyone who is interested in developing projects, ranging from projects that are very simple to those that potentially have worldwide application. The reader may be an entrepreneur who is ambitious to promote their ideas for any reason; or they may be a member of a development team in a company that is eager to fire up an NFC service. In either case, this book is well designed to satisfy every type of reader who is interested in writing any amount code on NFC.

WHAT THIS BOOK COVERS

Chapter 1 consists of introductory information on NFC technology. It gives some technical history and background information in NFC technology, and continues with the components of an NFC services setup. The component knowledge covers NFC-enabled mobile phones, NFC reader, NFC tags, and other complementary parts. This chapter will provide readers with enough knowledge on NFC at a macro level.

Chapter 2 contains the technical details of NFC technology that an NFC programmer will probably need. Details of NFC devices are initially covered in detail. This chapter consists of the technical details of three NFC operating modes, providing the reader with sufficient technical background, as well as the standards that must be followed when creating compatible programs within a development team. Details of the record types to be exchanged among NFC devices are included in the chapter for the same reason.

Chapter 3 and Chapter 4 consist of details of Android programming, for those who know Java but are not acquainted with Android in enough detail. The coverage of Android programming in this chapter is not extensive, but is enough to enable readers to continue with the later chapters on NFC programming using Android, as well as developing NFC applications further. All the necessary information on the Android development environment is included as well, to provide the necessary preparation for readers without knowledge about Android programming. Those who are already confident about Android programming can skip these chapters and proceed to the material that follows.

Chapter 5 and Chapter 6 contain instructions on how to program NFC in reader/writer mode, along with some examples. You should follow these instructions in sequence, in order to prevent overlooking important details.

Chapter 7 and Chapter 8 contain material on peer-to-peer mode, in a format similar to the previous two chapters.

Chapter 9 contains some information on card emulation (CE) mode. The details of CE mode are not covered in this book for two main reasons. The potential audience for CE mode seems very small when compared to the other modes. The standards of the other modes (i.e., reader/writer and peer-to-peer) are mostly well established, whereas finalized standards for CE mode programming on Android technology are still missing. Hence, it will be better to wait for the introduction of finalized CE programming standards for Android.

HOW THIS BOOK IS STRUCTURED

This book is structured in a top-down fashion. The chapters are isolated from each other, so that readers who have enough knowledge on the topic can just skip that chapter. The chapters are not integrated with each other in any way. The only exception is that dual chapters are created for the reader/writer (Chapter 5 and Chapter 6) and peer-to-peer (Chapter 7 and Chapter 8) operating modes — the earlier chapter explains how to program using the related mode, and the later one provides examples of that mode. Hence, the reader who does not have a complete understanding of programming using the related mode should read both chapters, while the reader who is confident about the programming of that mode can skip the earlier chapter and browse the later one containing the examples.

WHAT YOU NEED TO USE THIS BOOK

For NFC programming on Android, first you need to create an Android development environment. The most suitable way to do that is to install Android Development Tools (ADT) Bundle. ADT is available on Windows, MAC, and Linux operating systems. Moreover, in order to test NFC reader/writer mode applications, you need to have an NFC-enabled mobile phone and an NFC tag; in order to test NFC peer-to-peer mode applications, you need to have two NFC-enabled mobile phones; and in order to test card emulation mode applications, you need to have an additional Java Card that can be plugged-in to the mobile phone.

CONVENTIONS

To help you get the most from the text and keep track of what's happening, we've used a number of conventions throughout the book.

> **NOTE** *This is used for notes, tips, hints, tricks, or and asides to the current discussion.*

As for styles in the text:

➤ We *highlight* new terms and important words when we introduce them.

➤ We show keyboard strokes like this: Ctrl+A.

➤ We show filenames, URLs, and code within the text like so: `persistence.properties`.

➤ We present code in two different ways:

```
We use a monofont type with no highlighting for most code examples.
```

```
We use bold to emphasize code that is particularly important in the present context
or to show changes from a previous code snippet.
```

SOURCE CODE

As you work through the examples in this book, you may choose either to type in all the code manually, or to use the source code files that accompany the book. All the source code used in this book is available for download at `www.wrox.com`. Specifically for this book, the code download is on the Download Code tab at:

```
www.wrox.com/remtitle.cgi?isbn=1118380096
```

Throughout each chapter, you'll also find references to the names of code files as needed in listing titles and text.

Most of the code on `www.wrox.com` is compressed in a .ZIP, .RAR archive, or similar archive format appropriate to the platform. Once you download the code, just decompress it with an appropriate compression tool.

> **NOTE** *Because many books have similar titles, you may find it easiest to search by ISBN; this book's ISBN is 978-1-118-38009-3.*

Alternately, you can go to the main Wrox code download page at `www.wrox.com/dynamic/books/download.aspx` to see the code available for this book and all other Wrox books.

ERRATA

We make every effort to ensure that there are no errors in the text or in the code. However, no one is perfect, and mistakes do occur. If you find an error in one of our books, like a spelling mistake or faulty piece of code, we would be very grateful for your feedback. By sending in errata, you may save another reader hours of frustration, and at the same time, you will be helping us provide even higher quality information.

To find the errata page for this book, go to

```
www.wrox.com/remtitle.cgi?isbn=1118380096
```

And click the Errata link. On this page you can view all errata that has been submitted for this book and posted by Wrox editors.

If you don't spot "your" error on the Book Errata page, go to `www.wrox.com/contact/techsup port.shtml` and complete the form there to send us the error you have found. We'll check the information and, if appropriate, post a message to the book's errata page and fix the problem in subsequent editions of the book.

P2P.WROX.COM

For author and peer discussion, join the P2P forums at `http://p2p.wrox.com`. The forums are a Web-based system for you to post messages relating to Wrox books and related technologies and interact with other readers and technology users. The forums offer a subscription feature to e-mail you topics of interest of your choosing when new posts are made to the forums. Wrox authors, editors, other industry experts, and your fellow readers are present on these forums.

At `http://p2p.wrox.com`, you will find a number of different forums that will help you, not only as you read this book, but also as you develop your own applications. To join the forums, just follow these steps:

1. Go to `http://p2p.wrox.com` and click the Register link.

2. Read the terms of use and click Agree.

3. Complete the required information to join, as well as any optional information you wish to provide, and click Submit.

4. You will receive an e-mail with information describing how to verify your account and complete the joining process.

> **NOTE** *You can read messages in the forums without joining P2P, but in order to post your own messages, you must join.*

Once you join, you can post new messages and respond to messages other users post. You can read messages at any time on the Web. If you would like to have new messages from a particular forum e-mailed to you, click the Subscribe to this Forum icon by the forum name in the forum listing.

For more information about how to use the Wrox P2P, be sure to read the P2P FAQs for answers to questions about how the forum software works, as well as many common questions specific to P2P and Wrox books. To read the FAQs, click the FAQ link on any P2P page.

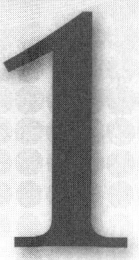

Overview of Near Field Communication

Currently, Near Field Communication (NFC) is one of the enablers for ubiquitous computing. This technology simplifies and secures interaction with the automation ubiquitously around you. Many applications you use daily such as credit cards, car keys, tickets, health cards, and hotel room access cards will presumably cease to exist because NFC-enabled mobile phones will provide all these functionalities.

The NFC ecosystem is designed from the synergy of several technologies, including wireless communications, mobile devices, mobile applications, and smart card technologies. Also, server-side programming, web and cloud services, and XML technologies contribute to the improvement and spread of NFC technology and its applications.

This chapter provides a brief background of the fundamentals and evolution of NFC technology. Then it gives a brief overview of NFC technology and the touching paradigm, including a comparison of NFC with other wireless technologies, and an introduction to smart NFC devices and operating modes with novel NFC applications in the industry.

UBIQUITOUS COMPUTING AND NFC

The history of modern computers comprises work that's been performed over the past 200 years. Personal computers (PCs) were an important step after early computers, changing the way that users interact with computers by using keyboards and monitors for input and output instead of primitive options such as punch cards and cables. The mouse also changed the way that humans interact with computers because it enables users to input spatial data in to a computer. Users became accustomed to using their hands to hold the mouse and pointing their fingers to click it. The movements of the pointing device are echoed on the screen by the movements of the cursor, creating a simple and intuitive way to navigate a computer's graphical user interface (GUI).

Touch screens changed the form of interaction even further and did so in a dramatic way. They removed the need for earlier input devices, and the interaction was performed by directly touching the screen, which became the new input device. In the meantime, mobile phones were introduced, initially for voice communication. Early forms of mobile phones contained a keypad. Those mobile phones with touch screens are considered to be state of the art because the screen is used for both input and output, which is more intuitive for users.

Ubiquitous computing is the highest level of interaction between humans and computers, in which computing devices are completely integrated into everyday life. Ubiquitous computing is a model in which humans do not design their activities according to the machines they need to use; instead, the machines are adjusted to human needs. Eventually, the primary aim is that humans using machines will not need to change their daily behaviors and will not even notice that they are performing activities with the help of machines.

As in modern computers and interfaces, increasing mobility of computing devices provided by mobile communications is also an important step in the development of ubiquitous computing capabilities and NFC. Mobile phones already had several communications options with the external environments before the introduction of NFC. When mobile phones were initially introduced, their primary goal was to enable voice communication. GSM (Global System for Mobile) communication further enabled functionality of mobile phones for several services, such as voice communication, short messaging service (SMS), multimedia message service (MMS), and Internet access. Also, the introduction of Global Positioning System (GPS) and Wireless Fidelity (WiFi) technologies (e.g., Infrared Data Association or IrDA) changed the way we use mobile phones. One communication option between mobile phones and computers was data transfer by USB — a physical port was used for this purpose, and cable was used for data transfer.

Later, Bluetooth technology was introduced, creating personal area networks that connect peripherals with computing devices such as mobile phones. Bluetooth became very popular in the early 2000s. Perhaps the most widely used function of Bluetooth is data exchange among mobile phones or between a mobile phone and another Bluetooth-enabled device such as a computer. Bluetooth enables communication among devices within a particular vicinity. However, secure data transfer cannot be performed completely with this technology because it is designed for wireless communication up to 10 meters, which allows malicious devices to alter the communication.

Currently, a new way of interacting has entered everyone's daily life: NFC technology can be identified as a combination of contactless identification and interconnection technologies. NFC operates between two devices in a short communication range via a touching paradigm. It requires touching two NFC-compatible devices together over a few centimeters. NFC communication occurs between an NFC mobile device on one side and an NFC tag (a passive RFID tag), an NFC reader, or an NFC mobile device on the other side. RFID is capable of accepting and transmitting beyond a few meters and has a wide range of uses. However, NFC is restricted for use within close proximity (up to a few centimeters) and also designed for secure data transfer. Currently, integration of NFC technology into mobile phones is considered a practical solution because almost everyone carries a mobile phone.

The main vision of NFC is the integration of personal and private information such as credit card or cash card data into the mobile phones. Therefore, security is the most important concern, and even the short wireless communication range provided by RFID technology is considered too long. Shielding is necessary to prevent unauthorized people from eavesdropping on private conversations because even nonpowered, passive tags still can be read over 10 meters. This is the point where NFC comes in.

NFC integrates RFID technology and contactless smart technologies within mobile phones. The evolution of NFC technology is illustrated in Figure 1-1. The gray areas in the figure indicate the technological developments that support the NFC environment directly. This chapter provides a brief overview of the technologies that make NFC evolution possible.

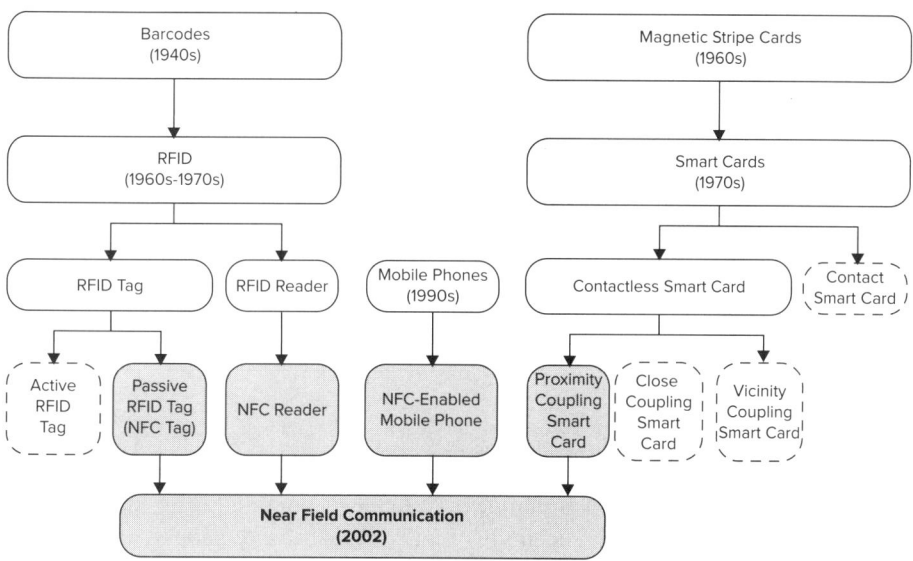

FIGURE 1-1

WIRELESS COMMUNICATION AS NFC

NFC technology also can be evaluated using a wireless communication aspect. Wireless communication refers to data transfer without using any cables. When communication is impossible or impractical through the use of cables, wireless communication is the solution. The range may vary from a few centimeters to many kilometers.

Wireless communication devices include various types of fixed, mobile, and portable two-way radios, cellular telephones, personal digital assistants, GPS units, wireless computer mice, keyboards and headsets, satellite television, and cordless telephones. Wireless communication allows communication without requiring a physical connection to the network.

Wireless communication introduces challenges that are somewhat harder to handle compared to wired communication; these challenges include interference, attenuation, unreliability, cost, and security. Wireless communication makes use of transmission of data over electromagnetic waves within the electromagnetic spectrum, as depicted in Figure 1-2.

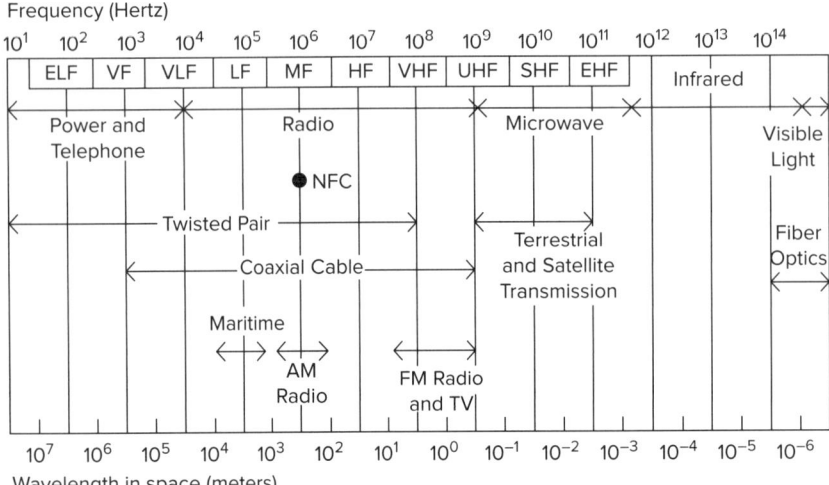

ELF: Extremely Low Frequency
VF : Voice Frequency
VLF: Very Low Frequency
LF : Low Frequency
MF : Medium Frequency

HF : High Frequency
VHF: Very High Frequency
UHF: Ultra High Frequency
SHF : Super High Frequency
EHF : Extremely High Frequency

FIGURE 1-2

The most straightforward benefit of wireless communication is mobility, which, indeed, has a big impact on everyone's daily life. Mobile communication supports not only the productivity and flexibility of organizations but also the social life of individuals because people can stay continuously connected to their social networks. Widely used wireless technologies include GSM, 3G, LTE (Long Term Evolution), Bluetooth, WiFi, WiMAX, and ZigBee.

Table 1-1 gives a brief summary and comparison of popular wireless technologies currently used around the world, according to their operating frequency, data rate, and operating range. GPRS, EDGE, and UMTS technologies represent wireless wide area networks (WWANs). Wireless local area networks (WLAN) follow these technologies with different frequencies and range, and then come the wireless personal area network (WPAN) technologies such as ZigBee and Bluetooth 2.0. NFC has the shortest communication range, which is followed by RFID technology.

TABLE 1-1: Overview of Some Wireless Technologies

WIRELESS TECHNOLOGY	OPERATING FREQUENCY	DATA RATE	OPERATING RANGE
UMTS	900, 1800, 1900 MHz	2 Mbps	Wide range
EDGE	900, 1800, 1900 MHz	160 Kbps	Wide range
GPRS	900, 1800, 1900 MHz	160 Kbps	Wide range
802.16 WiMAX	10–66 GHz	134 Mbps	1–3 miles
802.11b/g WiFi	2.4 GHz	54 Mbps	100 m
802.11a WiFi	5 GHz	54 Mbps	100 m
802.15.1 Bluetooth 2.0	2.4 GHz	3 Mbps	10 m
802.15.4 ZigBee	2.4 GHz	250 Kbps	70 m
NFC	13.56 MHz	106, 212, 424 Kbps	0–4 cm
RFID	125–134 kHz (LF)	1–200 Kbps	20 cm for passive
	13.56 MHz (HF)		400 cm for active
	400–930 MHz (UF)		
	2.5 GHz and 5 GHz		
	(microwave)		

RFID TECHNOLOGY

RFID is a wireless communication technology for exchanging data between an RFID reader and an electronic RFID tag through radio waves. These tags are traditionally attached to an object, mostly for the purposes of identification and tracking.

Figure 1-3 illustrates a simple RFID system and its components. The data transmission results from electromagnetic waves, which can have different ranges depending on the frequency and magnetic field. RFID readers can read data from, or write it to, tags.

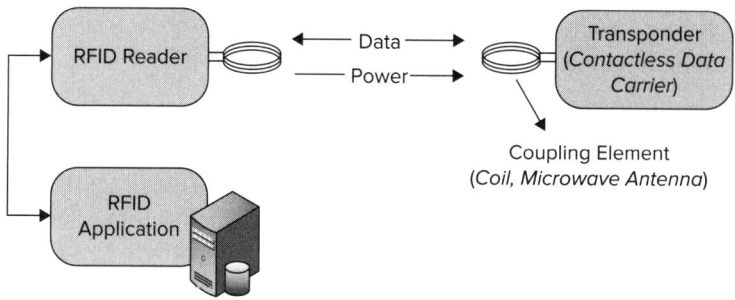

FIGURE 1-3

The connection between RFID readers and RFID applications uses wired or wireless networks in different sections of the communication. In the backend system, an RFID application is assigned specific information. RFID tags generally contain an integrated circuit (IC) and an antenna. The IC enables storing and processing data, modulating and demodulating radio frequency (RF) signals, and performing other functions. The antenna enables receiving and transmitting (reception and transmission of) the signal.

Essentials of an RFID System

An RFID system is made up of two major components: the transponder and reader. The transponder is a component that is located on a product or object to be identified, and the reader is a component that reads data from the transponder or writes to the transponder (as previously shown in Figure 1-3).

> **NOTE** *For more information on RFID systems, refer to* RFID Handbook: Fundamentals and Applications in Contactless Smart Cards, Radio Frequency Identification and Near-Field Communication *by Klaus Finkenzeller (Wiley, 2010).*

The transponder consists of a coupling element and an IC that carries the data to be transferred. The transponder is generally an RFID tag. RFID tags have a high capacity to store large amounts of data. They are divided into two major groups: passive tags, which have no power supply, and active tags, which have their own power supply. If the transponder is within the range of an RFID reader, it is powered by the incoming signal.

The reader typically contains a transceiver (high-frequency module) with a decoder for interpreting data, a control unit, and an antenna. Many RFID readers consist of an additional interface to send the received data to another system.

Common RFID Applications

RFID technology is being used all over the world for a wide variety of applications. Following are some examples:

➤ **Inventory systems:** Inventory tracking is a main area of RFID usage. RFID technology enables companies to manage inventory quickly and easily. It also enables companies to track reductions in out-of-stock items, increases in-product selling, as well as reductions in labor costs, simplification of business processes, and reduction of inventory inaccuracies.

➤ **Human implants:** Implantable RFID chips designed for animal-tagging are also being used in humans.

➤ **Animal identification:** Using RFID tags to identify animals is one of the oldest RFID applications. RFID provides identification management for large ranch operations and those with rough terrain, where tracking animals is difficult. An implantable variety of RFID tags located on animals is also used for animal identification.

➤ **Casino chip-tracking:** Some casinos are placing RFID tags on their high-value chips to track and detect counterfeit chips, observe and analyze betting habits of individual players, speed up chip tallies, and determine dealers' counting mistakes.

➤ **Hospital operating rooms:** An RFID reader and RFID-tagged disposable gauze, sponges, and towels are designed to improve patient safety and operational efficiency in hospitals.

SMART CARD TECHNOLOGY

A smart card includes an embedded IC that can be a memory unit with or without a secure microcontroller. It is a promising solution for efficient data storing, processing, and transfer and for providing a secure multiapplication environment. A typical smart card system contains smart cards, card readers, and a backend system. It may communicate with a reader using physical contact (contact smart card case) or a remote contactless RF interface (contactless smart card case). The reader connects to the backend system, which stores, processes, and manages the information.

> **NOTE** *For more information on smart cards, visit the website of Smart Card Alliance,* `www.smartcardalliance.org/`.

In terms of processing capability, smart cards are divided into two groups: memory-based and microprocessor-based. Memory-based smart cards can store data but need an external processing unit to do the processing. Smart cards with an embedded microcontroller can store large amounts of data and perform their own on-card functions, such as security-related operations and mutual authentication. These smart cards can interact intelligently with a smart card reader. These cards also have their own smart card operating system (SCOS). In terms of operating mechanisms, smart cards are divided into three groups: contact, contactless, and hybrid smart cards.

Types of Smart Cards: Capability-Based Classification

Smart cards are plastic cards with an embedded microprocessor and memory. Some smart cards have only nonprogrammable memory; thus, they have limited capabilities. Those smart cards with embedded or integrated microprocessors have various functionalities.

Memory-Based Smart Cards

Memory-based smart cards can store any kind of data, such as financial, personal, and other private information. However, they do not have any processing capability. These cards need to communicate with an external device such as a card reader using synchronous protocols to manipulate the data on the cards. These cards are widely used, for example, as prepaid telephone cards.

Microprocessor-Based Smart Cards

Microprocessor-based smart cards have on-card dynamic data processing capabilities. They have a microprocessor, as well as a memory. The microprocessor within the card manages the memory allocation and data management. Microprocessor-based smart cards are comparable with tiny computers, ones without an internal power source. These smart cards have an operating system (OS), namely SCOS, enabling you to manage the data on the smart card and allowing the smart cards to be multifunctional. They can store and process information and perform even complex calculations on the stored data. Unlike memory-based smart cards, they can record, modify, and process the data. Also, microprocessor-based smart cards have the capability to store large amounts of data when compared with memory cards.

SCOS

Until the end of the 1990s, it was very difficult to have more than one application running on a smart card due to the memory constraints of the IC chips. With the development of SCOSs, implementing several applications, running them simultaneously, and loading new ones during a card's active life became possible. Now, SCOSs enable more dynamic multiapplication platforms, and they are considered to be a really smart and powerful, secure computing environment for many new application domains.

Today each smart card has its own SCOS, which can be defined as a set of instructions embedded in the ROM of the smart card. Smart card architecture is depicted in Figure 1-4. The basic functions of SCOS include:

➤ Managing interchanges between a smart card and an external device such as a POS terminal

➤ Managing the data stored in memory

➤ Controlling the access to information and functions

➤ Managing security of the smart card, especially in terms of data integrity

➤ Managing the smart card's life cycle from its personalization to usage and termination

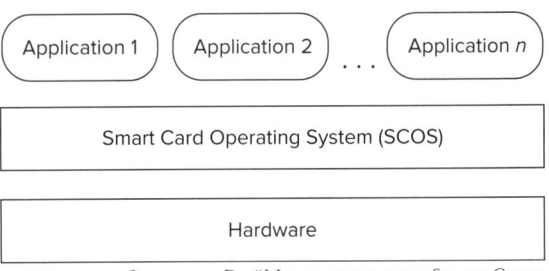

SAUVERON, D., "MULTIAPPLICATION SMART CARD: TOWARDS AN OPEN SMART CARD?," *INFORMATION SECURITY TECHNICAL REPORT*, 14(2), MAY 2009, 70-78

FIGURE 1-4

NOTE *For more information on SCOSs, refer to Damien Sauveron's article,"Multiapplication smart card: Towards an open smart card?"* (Information Security Technical Report, *Volume 14, Issue 2, May 2009, pages 70–78).*

Earlier in SCOS evolution, an application or a service on a smart card was written for a specific OS. Thus, the card issuer had to agree with a specific application developer as well as an operating system provider. This solution was costly and inflexible. Consumers needed to carry different smart cards for each service. Today the trend is toward an open operating system that supports multiple applications running on a single smart card. Currently, the most notable OSs that have bigger market exposure are MULTOS and JavaCard OS.

Types of Smart Cards: Mechanism-Based Classification

Smart cards are divided into three major groups in terms of the communication mechanism with outer devices: contact smart cards, contactless smart cards, and hybrid models.

Contact Smart Cards

Contact smart cards are embedded with a micro module containing a single silicon IC card that contains memory and a microprocessor. This IC card is a conductive contact plate placed on the surface of the smart card, which is typically gold plated. An external device provides a direct electrical connection to the conductive contact plate when the contact smart card is inserted into it. Transmission of commands, data, and card status information takes place over these physical contact points. Cards do not contain any embedded power source; hence, energy is supplied by the external device that the card currently interacts with. These external devices are used as a communications medium between the contact smart card and a host computer. These external devices can be computers, POS terminals, or mobile devices. Contact smart cards interacting with POS devices are typically used for payment purposes. Actually, the IC cards used on contact smart cards for payment purposes have the same hardware structure as those used in subscriber identity modules (SIMs) in mobile phones. They are just programmed differently.

The standards most related to contact smart cards are ISO/IEC 7810 and ISO/IEC 7816. They define the physical shape and characteristics of contact smart cards, electrical connector positions and shapes, electrical characteristics, communication protocols including commands exchanged with the cards, and basic functionality.

According to the ISO/IEC 7816 standard, the IC card has eight electrical gold-plated contact pads on its surface; they include VCC (power supply voltage), RST (reset the microprocessor), CLK (clock signal), GND (ground), VPP (programming or write voltage), and I/O (serial input/output line). Only the I/O and GND contacts are mandatory on a typical smart card; the others are optional. Two contacts (RFU) are reserved for future use (see Figure 1-5).

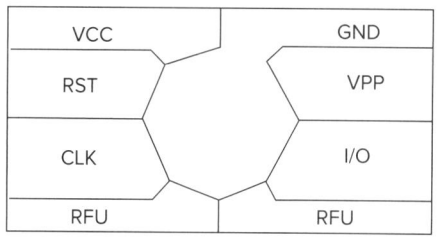

FIGURE 1-5

Contactless Smart Cards

A contactless smart card is a type of smart card that is processed without a need for physical contact with an external device. It is a combination of a microchip embedded within it and an antenna, which allows the card to be tracked (see Figure 1-6). Several wires form this antenna. In contactless smart cards, information is stored in the microchip, which has a secure microcontroller and internal memory. Unlike the contact smart card, the power supply to the contactless smart card is achieved with its embedded antenna. Data exchange between the smart card and an external device such as a smart card reader is performed with the help of this antenna. Electromagnetic fields for the card provide the power; hence, data exchange occurs between the card and the external device.

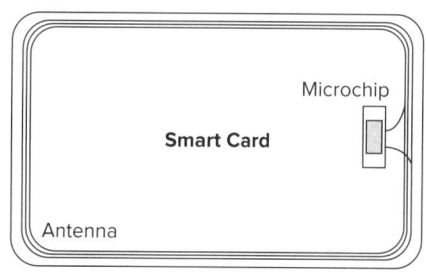

FIGURE 1-6

Contactless smart cards have the capability to store and manage data securely. They also provide access to the data stored on the card; they perform on-card functions such as enabling mutual authentication. They can easily and securely interact with a contactless external reader. The contactless communication can be performed only with devices in close proximity. Both the external device and contactless smart card have antennas, and they communicate using RF technology at data rates of 106–848 Kbps. As a contactless smart card is brought within the electromagnetic field of the card reader, the energy transfer starts from the card reader to the microchip on the smart card. The microchip is powered by the incoming signal from the card reader. After the microchip is powered, the wireless communication is established between the smart card and the card reader for data transfer.

Contactless smart card technology is used in applications for which private information such as health or identity data needs to be protected. It is also used in applications in which fast and secure transactions such as transit fare payment, electronic passports, and visa control are required. Contactless smart cards are often used for hands-free transactions. Applications using contactless smart cards must support many security features such as mutual authentication, strong information security through dynamic cryptographic keys, strong contactless device security, and individual information privacy. Contactless smart card technology is available in a variety of forms such as in plastic cards, watches, key fobs, documents, mobile phones, and other mobile devices.

Currently, three different major standards exist for contactless smart cards based on a broad classification range: ISO/IEC 10536 for close coupling contactless smart cards, ISO/IEC 14443 for proximity coupling smart cards, and ISO/IEC 15693 for vicinity contactless smart cards.

Hybrid Models

You might see other hybrid models of smart cards such as dual interface cards and hybrid cards. A dual interface card has both contact and contactless interfaces that contain only one chip. Such a model enables both the contact and contactless interfaces to access the same chip with a high level of security. A hybrid card contains two chips. One of those chips is used for a contact interface, and the other one is used for a contactless interface. These chips are independent and not connected.

Common Smart Card Applications

The first application of smart cards was prepaid telephone cards implemented in Europe in the mid-1980s. They were actually simple memory smart cards. Later, the application areas increased vastly. Today, some of the major application areas for microprocessor-based smart cards are finance, communications, identification, physical access control, transportation, loyalty, and healthcare. A smart card can even contain several applications.

NFC TECHNOLOGY

Philips and Sony jointly introduced NFC technology for contactless communications in late 2002. Europe's ECMA International adopted the technology as a standard in December 2002. The International Organization for Standardization (ISO) and the International Electrotechnical Commission (IEC) adopted NFC technology in December 2003. In 2004, Nokia, Philips, and Sony founded the NFC Forum to promote NFC technology and its services. NFC technology standards (see Table 1-2) are acknowledged by the International Organization for Standardization/

International Electrotechnical Commission (ISO/IEC), European Telecommunications Standards Institute (ETSI), and European Computer Manufacturers Association (ECMA).

TABLE 1-2: Major Standards for NFC Technology

STANDARDIZATION BODY	STANDARD	DESCRIPTION
ISO/IEC	ISO/IEC 18092	Near Field Communication Interface and Protocol (NFCIP-1)
	ISO/IEC 21481	Near Field Communication Interface and Protocol (NFCIP-2)
	ISO/IEC 28361	Near Field Communication Wired Interface (NFC-WI)
	ISO/IEC 14443	Contactless Proximity Smart Cards and their technical features
	ISO/IEC 15693	Contactless Vicinity Smart Cards and their technical features
ETSI	ETSI TS 102 190	Near Field Communication Interface and Protocol (NFCIP-1)
	ETSI TS 102 312	Near Field Communication Interface and Protocol (NFCIP-2)
	ETSI TS 102 541	Near Field Communication Wired Interface (NFC-WI)
	ETSI TS 102 613	Contactless front end (CLF) interface for UICC, physical and data link layer characteristics; Single Wire Protocol (SWP)
	ETSI TS 102 622	Contactless front end (CLF) interface for UICC, Host Controller Interface (HCI)
ECMA	ECMA 340	Near Field Communication Interface and Protocol (NFCIP-1)
	ECMA 352	Near Field Communication Interface and Protocol (NFCIP-2)
	ECMA 356	NFCIP-1 - RF Interface Test Methods
	ECMA 362	NFCIP-1 - Protocol Test Methods
	ECMA 373	Near Field Communication Wired Interface (NFC-WI)
	ECMA 385	NFC-SEC: NFCIP-1 Security Services and Protocol
	ECMA 386	NFC-SEC-01: NFC-SEC Cryptography Standard using ECDH and AES
	ECMA 390	Front-End Configuration Command for NFC-WI

continues

TABLE 1-2 *(continued)*

STANDARDIZATION BODY	STANDARD	DESCRIPTION
NFC Forum	NFC Digital Protocol Specification	Digital interface and the half-duplex transmission protocol of the NFC Forum Device
	NFC Activity Specification	Activities for setting up the communication protocol
	NFC Analog Specification	Analog interface of the NFC Forum Device
	NFC Controller Interface (NCI) Specification	NFC Controller Interface (NCI) between an NFC Controller (NFCC) and a Device Host (DH)
	Logical Link Control Protocol (LLCP) Specification	Supports P2P operation for NFC Applications
	NFC Data Exchange Format (NDEF) Specification	Common data format for devices and tags
	NFC Record Type Definition (RTD) Specification	Standard record types used in messages between devices/ tags
	Smart Poster RTD Specification	For posters with tags, text, audio, or other data
	Text RTD Specification	For records containing plaintext
	Uniform Resource Identifier (URI) Specification	For records that refer to an Internet resource
	NFC Types 1-4 Tag Operation Specifications	Defines NFC Forum Mandated Tag Types
	Connection Handover Specification	How to establish a connection with other wireless technologies

NFC is a bidirectional and short-range wireless communication technology that uses a 13.56 MHz signal with a bandwidth not more than 424 Kbps. NFC technology requires touching two NFC-compatible devices together over a few centimeters.

> **NOTE** *For more information on NFC technology and its ecosystem, visit the* *NFC Forum website:* www.nfc-forum.org/.

User awareness is essential to perform NFC communication. The user first interacts with a smart object such as an NFC tag, NFC reader, or another NFC-enabled mobile phone using a mobile phone (see Figure 1-7). After touching occurs, the mobile device may make use of received data and thus may additionally use mobile services as well, such as opening a web page or making a web service connection.

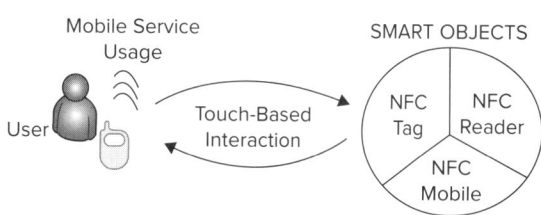

FIGURE 1-7

Depending on the mentioned interaction styles, NFC technology operates in three operating modes: reader/writer, peer-to-peer, and card emulation. Each operating mode uses its specific communication interfaces (ISO/IEC 14443, FeliCa, NFCIP-1 interfaces) on the RF layer as well as having different technical, operational, and design requirements that are explicitly presented and illustrated in Chapter 2, "NFC Essentials for Application Developers."

NFC Devices

NFC technology uses the following smart devices:

➤ **NFC-enabled mobile phone:** NFC-enabled mobile phones, which also are referred to as NFC mobiles, are the most important NFC devices. Currently, integration of NFC technology with mobile phones (thereafter introducing NFC-enabled mobile phones) creates a big opportunity for the ease of use, acceptance, and spread of the NFC ecosystem.

➤ **NFC reader:** An NFC reader is capable of data transfer with another NFC component. The most common example is the contactless point of sale (POS) terminal, which can perform contactless NFC-enabled payments when an NFC device is touched against the NFC reader.

➤ **NFC tag:** An NFC tag is actually an RFID tag that has no integrated power source.

NFC works in an intuitive way. The touching action is taken as the triggering condition for NFC communication. Two NFC devices immediately start their communication when they are touched. The NFC application is designed so that when the mobile touches some other NFC component that contains the expected form of data, it boots up. Hence, the user does not need to interact with the mobile device anymore but just touches one appropriate NFC device, which may be an NFC tag, an NFC reader, or another NFC-enabled mobile phone, because the coupling occurs intuitively and immediately. When you consider ubiquitous computing requirements, this is a useful property of NFC communication.

For each NFC communication session, the party who initiates the communication is called the *initiator*, whereas the device that responds to the requests of the initiator is called the *target*. This case is analogous to the well-known client/server architecture. Table 1-3 shows the possible interaction styles of NFC devices in terms of initiator and target roles.

TABLE 1-3: Interaction Styles of NFC Devices

INITIATOR	TARGET
NFC Mobile	NFC Tag
NFC Mobile	NFC Mobile
NFC Reader	NFC Mobile

In an active/passive device approach, if an NFC device has an embedded power source, it can generate its own RF field and naturally initiates and may lead communication. This device is called an *active device*. On the other hand, if it does not have any embedded power source, it can only respond to the active device; in this case, it is called a *passive device*.

An initiator always needs to be an active device because it requires a power source to initiate the communication. The target, however, may be either an active or a passive device. If the target is an active device, it uses its own power source to respond; if it is a passive device, it uses the energy created by the electromagnetic field, which is generated by the initiator that is always an active device.

Consider an NFC tag, which is a low-cost and low-capacity device. It does not contain any power source and needs an external power source to perform any activity. Thus, an NFC tag is always a passive device and always a target because it does not include any energy source by design. It stores data that an active device can read.

NFC Operating Modes

As mentioned previously, three existing NFC operating modes are the reader/writer, peer-to-peer, and card emulation modes with different interaction styles. The reader/writer mode enables NFC-enabled mobile devices to exchange data with NFC tags. The peer-to-peer mode enables two NFC-enabled mobiles devices to exchange data with each other. In the card emulation mode, the user interacts with an NFC reader to use a mobile phone as a smart card, such as a contactless credit card.

Service usage in each NFC operating mode differs because the interacted smart objects are different and provide distinct usage scenarios. Each operating mode has its own characteristics; therefore, it is possible to define a usage model for each operating mode. Generic usage models define the mandatory characteristics of each operating mode, along with the usage principle of the technology. The following subsections describe each operating mode and its generic usage model in detail.

Reader/Writer Operating Mode

The reader/writer mode is about the communication of an NFC-enabled mobile phone with an NFC tag for the purpose of either reading data from or writing it to those tags. This mode internally defines two different modes as reader mode and writer mode.

In reader mode, the initiator reads data from a 13.56 MHz tag or NFC tag, which consists of the requested data. The specification of the NFC tags is defined by the NFC Forum; therefore, the NFC tags may also be called NFC Forum–mandated NFC tags. The NFC tag mentioned here is one of the NFC Forum–mandated tag types as described in Chapter 2. In addition to the requirement that the NFC tag already consists of the requested data, it also consists of the program that returns the requested data to the initiator.

In writer mode, the mobile phone acts as the initiator and writes data to the tag. If the tag already contains any data prior to the writing process, it is overwritten. The algorithm may even be designed so that the initiator will update by modifying the previously existing data instead of overwriting it. Although it is not a common option, an NFC reader, in addition to the mobile phone, may also be used to read data from a tag. You may even envisage an NFC reader that writes data to an NFC tag. The only possible data rate in this mode is 106 Kbps. A schematic representation of the reader/ writer mode is given in Figure 1-8.

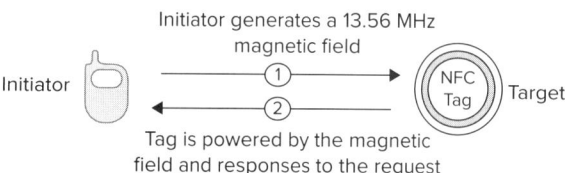

FIGURE 1-8

A mobile phone can perform several actions after it reads the data from the tag. If the tag stores a URL, the phone can launch a web browser and display the received web page. The features of mobile phones, such as processing power, audio/video capability, and Internet access, provide many opportunities for both users and service providers when reader/writer mode is used. Applications in this mode are countless and can be very innovative.

Generic Usage Model of Reader/Writer Mode

The generic usage model of the reader/writer operating mode is explained here step by step and also illustrated in Figure 1-9:

1. **Read request:** A user first requests data by touching a mobile phone to an NFC tag, which can be embedded in various components such as a smart poster or product package.

2. **Data transfer:** The data that resides in the tag is transferred to the mobile phone.

3. **Processing within device:** When data is transferred to the mobile phone, it can be used for several purposes, such as pushing an application, displaying data to the user, or processing data by an application for additional purposes.

4. **Additional service usage:** This optional step takes advantage of the mobile phone's advanced capabilities and mostly involves Internet connectivity capability. When data is processed in the mobile phone, it can be used for further operations via the Internet such as connecting to a service provider by using an application's web service.

5. **Write request:** The user requests the capability to write data to an NFC tag by touching a mobile phone to it.

6. **Acknowledgment:** The NFC tag replies with the acknowledgment data, informing the user about the success of the operation.

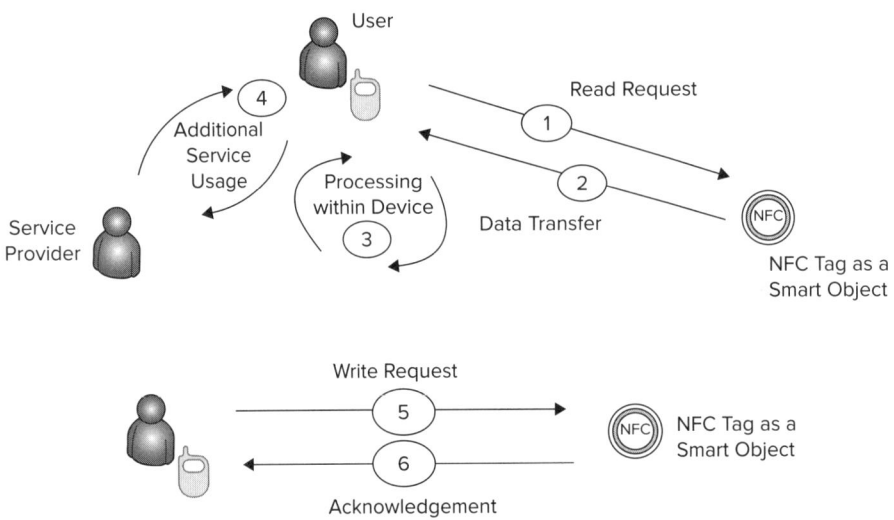

FIGURE 1-9

Peer-to-Peer Operating Mode

Peer-to-peer mode enables two NFC-enabled mobile devices to exchange information such as a contact record, a text message, or any other kind of data. This mode has two standardized options: NFC Interface and Protocol-1 (NFCIP-1) and Logical Link Control Protocol (LLCP, which is used on the top of NFCIP-1). They are described further in Chapter 2.

NFCIP-1 takes advantage of the initiator-target paradigm in which the initiator and the target devices are defined prior to starting the communication. However, the devices are identical in LLCP communication. After the initial handshake, the application running in the application layer makes the decision. Because of the embedded power to mobile phones, both devices are in active mode during the communication in peer-to-peer mode. Data is sent over a bidirectional half-duplex channel, meaning that when one device is transmitting, the other one has to listen and should start to transmit data after the first one finishes. The possible data rates in this mode are 106, 212, and 424 Kbps. A schematic representation of the peer-to-peer mode is given in Figure 1-10.

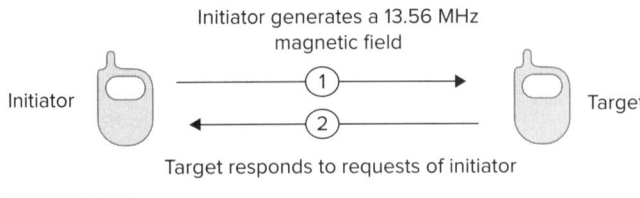

FIGURE 1-10

Generic Usage Model of Peer-to-Peer Mode

In peer-to-peer mode, users communicate with each other using NFC-enabled mobile phones. In the simplest option of this mode, no service provider is used. If users intend to use any services on the Internet, a service provider also may be included in the process. The generic usage model of peer-to-peer operating mode is explained here step by step, and Figure 1-11 shows each step of the generic usage model:

1. **Data request/transfer:** Two users exchange data via mobile phones.

2. **Additional service usage:** When data is shared between mobile phones, this data can optionally be used for additional purposes such as saving a received business card to a database over the Internet after a successful share or starting a friendship on a social network.

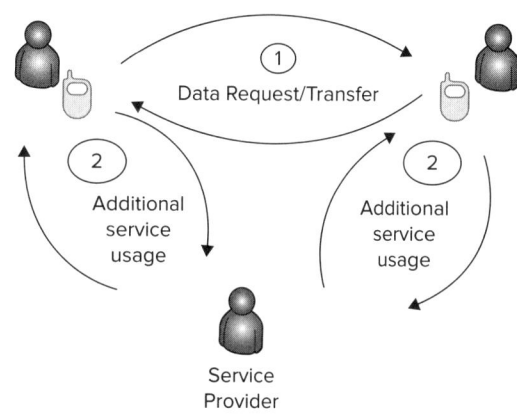

FIGURE 1-11

Card Emulation Operating Mode

Card emulation mode provides the opportunity for an NFC-enabled mobile device to function as a contactless smart card. A mobile device can even store multiple contactless smart card applications on the same the smart card. The most implemented examples of emulated contactless smart cards are credit cards, debit cards, and loyalty cards.

In this operating mode, an NFC-enabled mobile phone does not generate its own RF field; the NFC reader creates this field instead. This behavior is surprising because the mobile is an active device and therefore can use its own energy. Currently supported communication interfaces for the card emulation mode are: ISO/IEC 14443 Type A and Type B, and FeliCa, which are described in Chapter 2.

Card emulation mode is important because it enables payment and ticketing applications. It is also practical because it is compatible with the existing smart card infrastructure. A schematic representation of card emulation mode is given in Figure 1-12.

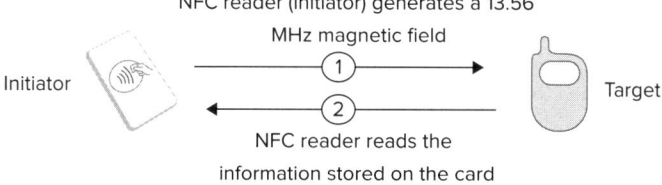

FIGURE 1-12

Generic Usage Model of Card Emulation Mode

In card emulation mode, the user interacts with an NFC reader, traditionally using a mobile phone as a smart card. The NFC reader is owned by a service provider, which is possibly connected to the Internet. The user connects to a service provider through an NFC reader possibly without notifying the service provider. The generic usage model of the card emulation operating mode is explained here step by step. Figure 1-13 illustrates the steps in the generic usage model of card emulation mode:

1. **Service request:** The user makes a request to a service provider by touching a mobile phone to an NFC reader. Required data is transferred from the mobile phone to the service provider through the NFC reader.

2. **Background services:** The service provider runs required backend services after getting the required data from the user's mobile device. Examples of these services are credit card authorization and ticket validation.

3. **Service usage + data (optional):** The service provider returns a service to the user, such as issuing a ticket that has already been purchased using the payment card or authorizing the payment.

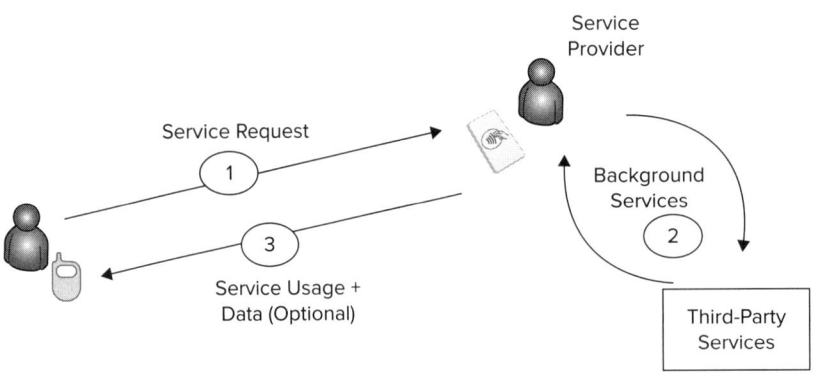

FIGURE 1-13

NFC Applications

NFC operating modes have different characteristics so that each mode provides different use cases. The following sections describe some novel NFC applications depending on their operating modes.

Reader/Writer Mode Applications

In reader/writer mode, numerous use case opportunities are available. Transferred data can be text, a URL, a product identification, or some other type of data (see Figure 1-14). After the transfer operation, the mobile phone can use the data for many purposes according to the design of the use case.

FIGURE 1-14

Companies or professionals might design projects in this mode because of the flexibility of different data types to be stored on the NFC tag, as well as the flexibility regarding how to use that information. Hence, a wide range of applications in health, education, location-based services, remote services, social networking, and entertainment potentially may be generated using the reader/writer mode.

Peer-to-Peer Mode Applications

As mentioned previously, peer-to-peer mode provides easy data exchange between two NFC mobiles. Easy data exchange between two NFC-compatible devices with high-performance computing capabilities provides the possibility for secure exchange of private data that makes use of the processing power (see Figure 1-15). NFC devices can transfer data in a few centimeters, so exchanging private and important data may be one of the key future applications of this mode. Pairing Bluetooth devices; exchanging business cards or other data such as file, image, and text messages; making new friends on online networks (Twitter, Facebook, and so on); and gaming are examples of possible implementations using this mode.

FIGURE 1-15

Card Emulation Mode Applications

Following are some examples illustrating how you can implement card emulation mode in various applications:

➤ **Payment:** You can have different types of NFC payment applications. There is no doubt that the most important payment options are credit and debit card applications, which can be triggered by NFC readers. Other opportunities for NFC payment include storing and using vouchers, using gift cards, and so on (see Figure 1-16).

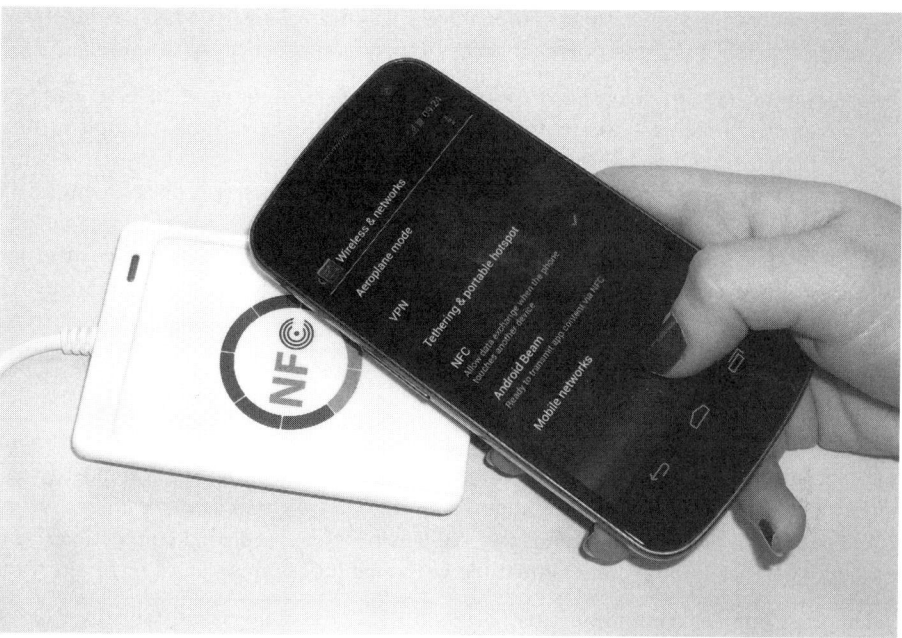

FIGURE 1-16

➤ **Loyalty:** Users can gain loyalty points at payment points and later can use them to shop free or obtain a gift. Also, NFC readers using this operating mode can use coupons downloaded via smart posters using reader/writer mode.

➤ **Ticketing:** Ticketing use cases can be implemented in different forms. Users can store different types of tickets such as theater, bus, and flight tickets previously downloaded via smart posters or by some other method. Users can then use these tickets in turnstiles or ticketing points via the card emulation mode. They can also store and validate prepaid or monthly ticketing cards.

➤ **Access control:** Access control use cases enable users to embed their access control objects in their mobile devices. Examples of these cases include electronic keys for cars, buildings, secure areas, and hotel rooms. Hotel check-in is an interesting use case that enables users to receive a room key via OTA technology prior to arriving at the hotel and enabling the users to directly check in to the room. Thus, there is no reason to stop at the reception area on arrival in this instance.

➤ **Identity services:** Another interesting use case of this mode is storing identity-based information on mobile devices and enabling authorized personnel to access it. An example of this type of service is storing patient data. A patient's medical history can be stored on a mobile device, and the user can then choose to give permission to a doctor to access that data via an NFC reader. The medical data need not to be stored on the hospital's or medical insurance company's servers. This use case increases user privacy because undesired third parties such as insurance companies are not able to access the patient's information. Even more interesting applications can be developed in this category, such as integrating national identification cards, passports, fingerprints, and driver's licenses to mobile phones.

➤ **Smart environment:** The smart environment case refers to using NFC technology in smart environments such as a smart home or office. The most common example is managing smart environments via preconfigured data on a mobile phone. In this case, when a user enters a smart environment, the user can adjust specific smart environment settings such as brightness level and music-melody selection through a mobile device. The device can also be integrated with an access control mechanism so that when the user opens the door using an NFC electronic key, that user can activate a personalized smart environment.

SUMMARY

Ubiquitous computing refers to the next level of interaction between humans and computers in which computing devices are completely integrated into everyday life. NFC is mostly regarded as an important step toward ubiquitous computing. NFC uses the touching paradigm for interaction. Users need to touch their mobile phones to a reader or tag to establish a connection. NFC is an extension of RFID technology and compatible with contactless smart card technology interfaces.

NFC technology enables communication based on RFID technology and ISO/IEC 14443 infrastructures. It operates in three modes (reader/writer, peer-to-peer, and card emulation) with an RF of 13.56 MHz, where communication occurs between a mobile phone with NFC capability on one side and, on the other side, an NFC tag, an NFC mobile, or an NFC reader.

NFC technology allows people to integrate their daily-use cards such as loyalty and credit cards into their mobile phones. In addition to integrating daily-use cards into mobile devices, NFC technology brings innovations to mobile communications. It enables users to easily communicate or exchange data simply by touching two mobile phones to each other. Moreover, NFC technology gives NFC reader capability to mobile phones; hence, they can read RFID tags. Mobile phones' increasing processing power, Internet access, and many more features gain advantage from this reader functionality and open ways for new and innovative services.

Many NFC trials and applications have been conducted around the world. Generally, all trials concluded that with the development of NFC technology, mobile phones are likely to become safer, speedier, and more convenient and more fashionable physical instruments.

It is true that NFC technology brings simplicity to transactions, provides easy content delivery, and enables information sharing. At the same time, it builds new opportunities for various stakeholders — for example, providing mobile operators, banks, transport operators, and merchants with faster transactions, less cash handling, and new operator services.

NFC Essentials for Application Developers

➤ Basic architecture and main components of NFC mobile

➤ Introduction to Secure Element (SE) and its alternatives

➤ Importance of over-the-air (OTA) technology

➤ Standards used by NFC: NFCIP-1, NFCIP-2, and proximity contactless smart card standards

➤ Communication architectures and essentials of NFC operating modes

➤ NFC Forum tag types: Type 1, Type 2, Type 3, and Type 4

➤ NFC Data Exchange Format (NDEF) and record types

➤ Logical Link Control Protocol (LLCP) for link-level communication

➤ Evaluation of NFC service platforms diversity

This chapter introduces and elaborates on NFC technical essentials. It initially introduces the NFC mobile's architecture and its main components: Secure Elements (SEs), NFC controller, Single Wire Protocol (SWP), NFC Wired Interface (NFC-WI), and so on. SE is the most important component of the NFC mobile because it guarantees users and service providers that transactions take place in a protected environment of NFC mobiles. The chapter also presents various SE options and briefly explains the importance of over-the-air (OTA) technology in management of SEs.

This chapter also focuses on the communication architecture of operating modes (reader/writer, peer-to-peer, and card emulation operating modes) and protocols and standards in

NFC. The NFC Forum provides valuable specifications and standards especially for the communication infrastructure of the reader/writer and peer-to-peer operating modes. Card emulation mode is quite different from the others; it implicitly gives the smart card operating capability to an NFC device that has a secure environment.

NFC MOBILE

A mobile device integrated with NFC technology is typically composed of various integrated circuits, SEs, and an NFC interface (see Figure 2-1). The NFC interface is composed of a contactless analog/digital front end called an NFC Contactless Front End (NFC CLF), an NFC antenna, and an integrated circuit called an NFC controller to enable NFC transactions. In addition to an NFC controller, an NFC-enabled mobile phone has at least one SE, which is connected to the NFC controller for performing secure proximity transactions with external NFC devices. The SE provides a dynamic and secure environment for programs and data. It enables secure storage of valuable and private data such as a user's credit card information, and secure execution of NFC-enabled services such as contactless payments. Also, more than one SE can be integrated into an NFC mobile.

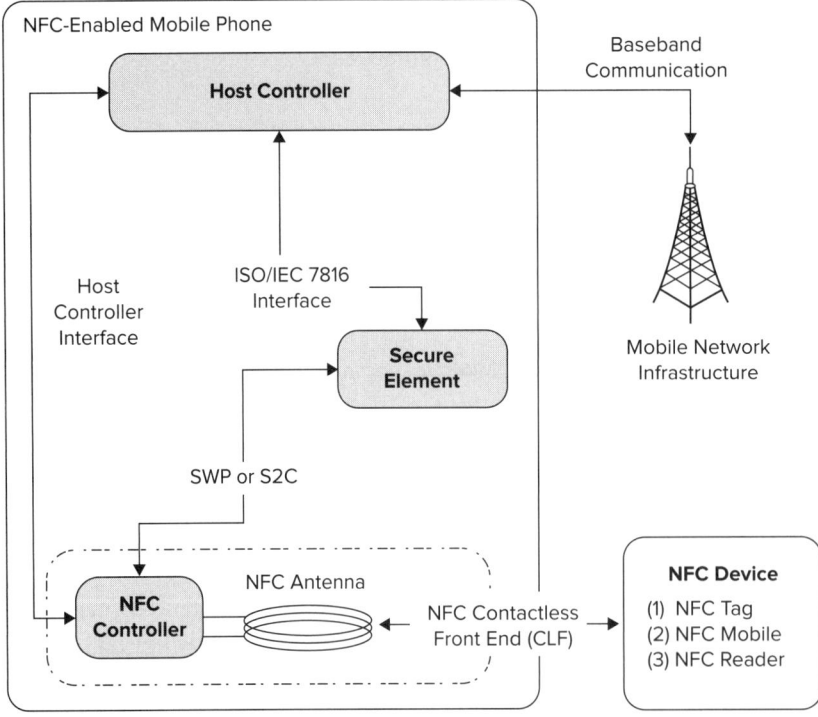

FIGURE 2-1

The two common interfaces currently supported between SEs and the NFC controller are the SWP and NFC-WI. The SE can be accessed and controlled internally from the host controller as well as externally from the RF field. The host controller (baseband controller) is the heart of any mobile phone. A host controller interface (HCI) creates a bridge between the NFC controller and host controller. An ISO/IEC 7816 interface supports the linkage of SEs to the host controller. The host controller sets the operating modes of the NFC controller through the HCI, processes data that is sent and received, and establishes a connection between the NFC controller and the SE. Furthermore, it maintains the communication interface, peripherals, and user interface.

SE

NFC-enabled services must reassure users and service providers that the transaction takes place in a protected environment. This protection is achieved by using an SE that provides the security required to support various business models. The SE is a combination of hardware, software, interfaces, and protocols embedded in a mobile handset that enables secure storage and processing.

An SE needs to have an operating system as usual. An operating system, such as MULTOS or JavaCard OS, on a mobile device supports the secure execution of applications and secure storage of application data. The operating system may also support the secure loading of applications. If NFC-enabled applications are saved and executed in the memory of the NFC-enabled mobile phone's host controller, these applications are not protected against unintentional deletion or intentional manipulation of the saved data in memory. Applications transmit data only between NFC-enabled mobile phones or collect information from smart posters.

In contactless ticketing, payment, and other similar application cases, security is an important and nontrivial issue. These applications use valuable data, so the storage of valuable, private information such as credit card information in unsecured memory is unacceptable. The data could be transmitted via a GSM (Global System for Mobile Communication) interface to a third party who may misuse the information in such a case.

To solve this issue, relevant NFC applications need to be executed and saved in the memory of an SE of the NFC-enabled mobile phone. A variety of modules can serve as SEs, such as stickers, Universal Integrated Circuit Cards (UICCs), memory cards, and embedded hardware. An SE is necessary for various applications such as payment, ticketing, government, and other applications for which secure authentication and a trusted environment are among the prerequisites.

SE Alternatives

Up to now, various SE alternatives that have entered the market can enable financial institutions and other companies to offer secure NFC-enabled services and empower the NFC ecosystem takeoff. Mainly, SE options can be grouped as removable SEs, nonremovable SEs, software-based SEs on dedicated hardware, and other flexible SE solutions. Actually understanding the characteristics of these SEs plays a significant role for different stakeholders and pricing models in the NFC value chain. The dominating SE will have a strong position to build trusted services on it. Figure 2-2 shows the SE alternatives currently possible within the NFC ecosystem for each category of SE.

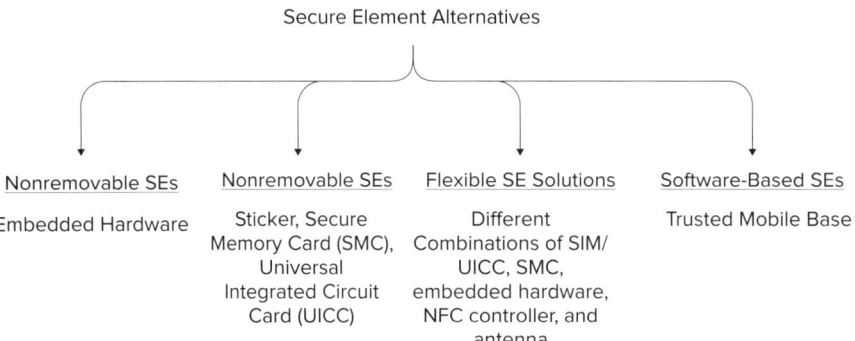

FIGURE 2-2

Embedded Hardware

An embedded hardware-based SE alternative is a smart card integrated with the mobile phone, and it cannot be removed after production. Thus, the level of security provided by this SE is as high as the one supported by a smart card. This chip is embedded into the mobile phone during the manufacturing stage and must be personalized after the device is delivered to the end user. Due to its embedded nature on NFC mobiles, an SE of this type obviously cannot be used in other mobile phones; it needs to be replaced and personalized every time the mobile phone is used by any other user.

Stickers

Stickers are actually somewhat different from the other two removable SE options: Secure Memory Card (SMC) and Universal Integrated Circuit Card (UICC). The purpose of NFC stickers is to allow service providers to launch pilots quickly and enable deployment of NFC services such as payment, loyalty, and transportation. Stickers are typically contactless smart cards or tags designed to be attached on the back of mobile phones. They have a specifically designed NFC antenna combined with a ferrite-backing layer to cut distortion to and from the mobile phone's components and its radio signal.

The two types of stickers are active and passive. Active stickers are connected to a mobile phone's application execution environment, namely its operating system, which makes the stickers become an integrated part of the mobile. Passive stickers do not allow dynamic application management because they do not have any cabled connection with the mobile. Active stickers enable all NFC services operating in reader/writer, peer-to-peer, and card emulation mode; also, OTA provisioning and life-cycle management of active stickers are easy because of their connection with the mobile phones.

SMC

An SMC is made up of memory, an embedded smart card element, and a smart card controller. Thereby, an SMC provides the same high level of security as a smart card, and it is compliant with most of the main standards and interfaces of smart cards such as GlobalPlatform, ISO/IEC 7816, and JavaCard. With the removable property and a large capacity memory, an SMC can host a large number of applications in it. It does not need to be reissued even when the customer buys a new NFC mobile. The user can simply insert it into the new device. An SMC can be inserted in any device supporting NFC technology and is not limited to use by one mobile phone. It can be issued for extended usage across a user's electronic devices such as laptops or portable media players.

UICC

UICC is a generic multiapplication platform for smart card applications on which SIM (Subscriber Identity Module) and USIM (Universal Subscriber Identity Module) are implemented. It was standardized by the ETSI Project Smart Card Platform with the aim of defining a physical and logical platform for all smart card applications and thus to enable development of advanced security methods, especially for financial transactions. UICC provides an ideal environment for NFC applications that is personal, secure, portable, and easily managed remotely via OTA technology.

UICC-based SEs are to be issued by a party who is usually MNO (Mobile Network Operator). Mainly, it hosts some required applications from a UICC issuer such as SIM or USIM (UMTS/3G SIM) applications to authenticate a user in a 3G network. In addition to SIM and USIM applications, UICC can host nontelecom applications such as loyalty, ticketing, healthcare, access control, and ID applications from various service providers.

Currently, GlobalPlatform provides the most promising standards for management of multiple applications on the same UICC smart cards. In accordance with those standards, UICC-based SEs provide separate security domains with a secret administrative key for each application. The card's OS implements a firewall that tries to prevent applications from accessing or sharing data between them. However, some issues remain unsolved on UICC card management in NFC-based services.

Flexible SE Solutions

Because of a massive lack of NFC mobiles within the market, especially until recently, several alternative architectures based on different communication options of SEs and NFC have been proposed. SE manufacturers are performing some hardware improvements on the existing SE solutions, adding NFC functionalities to those SEs, and providing mobile phone device and hardware-independent alternatives. Especially SMC- and SIM-based SEs with built-in NFC antennas have great potential as NFC bridge devices; they shorten the time to market for contactless payment and similar applications. These solutions seem to be a good way of promoting NFC applications; however, they are not persistent solutions. Instead, they are mostly provided for specific NFC services. Some examples from the industry are integration of NFC with SIM cards (SIM Application Toolkit); standard SIM cards with only NFC antenna; standard SD or microSD cards hosting NFC chips, antennas, and SEs; and standard SD or microSD cards hosting only NFC antennas and SEs.

TMB

Trusted Mobile Base (TMB) is a promising upcoming technology that is hosted at the root of mobile phones. It is defined as a secure isolated section on the core processors (CPU) of mobile phones. Various secure NFC-enabled applications can be provided flexibly via OTA technology. Currently, TMB is not available, but according to Mobey Forum, it has the full potential of becoming an SE in the future.

TMBs can enable secure user interfaces and OTA-provisioning services to security domains. TMBs can also be identified as "open platforms" with standardized interfaces and assist in achieving a required security level together with the other SE alternatives. Because TMBs are built into the CPU, they have no additional hardware costs. TMB-related services can be provided via OTA similar to other SE alternatives.

> **NOTE** *For more information on SEs, refer to Mobey Forum,*
> *Alternatives for banks to offer secure mobile payments, White Paper,*
> *Version 1,* www.mobeyforum.org/Press-Documents/Press-Releases/
> Alternatives-for-Banks-to-offer-Secure-Mobile-Payments.

SE Management

Today, MNOs have the ability to remotely configure mobile devices; upload and install new applications based on user requests and approval; upload and execute application software and OS updates; remotely lock a device to protect the application and application data when the mobile is, for example, lost or stolen; troubleshoot the device remotely; and perform additional services similar to OTA technology.

OTA technology also contributes to the dynamic spirit of NFC-based system adaptability to flexible environments. OTA technology enables loading and installation of new NFC applications on SEs, especially remotely on UICCs; activation and deactivation of SEs; remote service management (installation, personalization, update, termination); and life-cycle management (card block, unblock, re-issuance, PIN reset, change, parameter updates) of NFC applications on SEs and other online services.

High-capacity bearers — those being used in OTA technology — are important in providing an NFC solution (see Figure 2-3). For example, embedded and SMC-based SEs can be accessed via OTA only through a MIDlet proxy. This connection requires that the communication is initiated on the mobile handset side and a secure HTTP connection is established using 3G communication. Several kilobytes' worth of data needs to be transferred to the UICC-based SE when downloading activation application data or an NFC application itself. For example, when you use GPRS/UMTS and Bearer Independent Protocol (BIP), applications are rapidly deployed over the air to the UICC.

NFC Interface

The NFC interface is composed of a contactless analog-to-digital front end that is called an NFC contactless front end (NFC CLF), an NFC antenna, and an IC called an NFC controller to enable NFC transactions, as shown in Figure 2-4.

The NFC controller enables the establishment of the NFC link in a mobile phone. It works as a modulator and demodulator between the analog RF signal and NFC antenna. The NFC controller supports both active and passive communication with various modulation types. Typically, an NFC controller is compliant with the NFCIP-1 (Near Field Communication Interface and Protocol - 1) protocol (peer-to-peer mode), as well as the other two operating modes (reader/writer and card emulation modes). Also, other RFID protocols such as ISO/IEC 15693 are often supported.

NFC CLF is the analog front end of the NFC controller. The NFC CLF logical interface defines the protocol on top of the data link layer, as well as how the messages are transmitted between the SE and the NFC CLF. It is theoretically independent from the underlying interface (physical and data link interface), which carries the messages.

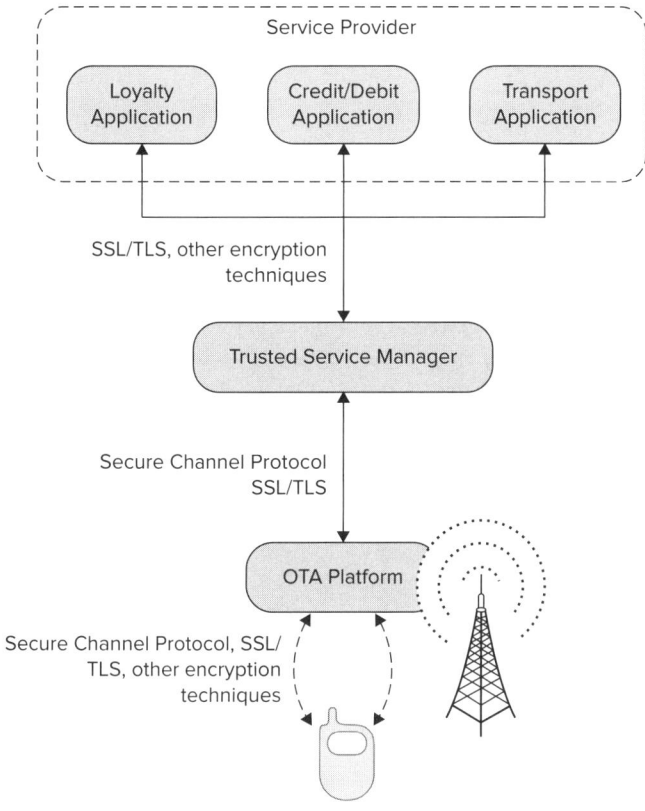

FIGURE 2-3

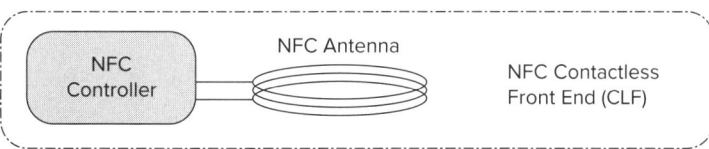

FIGURE 2-4

All SE design models have an interface between the SE and NFC controller and also between the host controller and NFC controller. The data transmitted via the contactless interface is directly forwarded by the NFC controller to the SE and vice versa. The host, which is the nonsecure part of the system, is not involved in the transaction.

Interface Between SE and NFC Controller

Various technical options exist for designing the interface between the SE and NFC controller. The most promising two options are: NFC WI and SWP. The most important difference between them is that SWP uses one physical line, whereas NFC-WI uses two lines. It is worth mentioning that these two options are not alternatives to each other but options to be used in certain and appropriate places instead (e.g., NFC-WI is more for embedded SE-based mobiles while SWP is for UICC-based mobiles).

NFC-WI

NFC-WI (also called S2C) is a digital wire interface standardized by ECMA 373, ISO/IEC 28361, and ETSI TS 102 541. Three transmission rates supported by NFC-WI are: 106, 212 and 424 Kbps. The SE is defined as a transceiver, and the NFC controller is defined as a front end in this protocol. The SE is connected to the NFC controller via two wires. NFC-WI defines the Signal-In (SIGIN) and Signal-Out (SIGOUT) wires between the transceiver and front end, as illustrated in Figure 2-5. The transceiver is the entity that drives the SIGIN wire and receives on the SIGOUT wire. The front end is the entity that drives the SIGOUT wire and receives on the SIGIN wire.

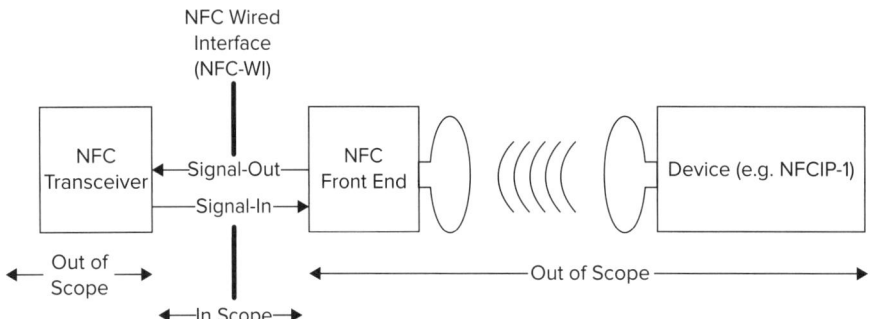

ECMA INTERNATIONAL (2006), ECMA 373: NEAR FIELD COMMUNICATION WIRED INTERFACE (NFC-WI), JUNE 2006.

FIGURE 2-5

This digital wire interface carries two binary signals that are defined as HIGH and LOW. Both of them transmit modulation signals between the NFC controller and SE and are digitally received and sent by the RF interface. The transceiver drives the SIGIN wire with a binary signal of either HIGH or LOW. The front end receives the binary signal that is on the SIGIN wire, and it drives the SIGOUT wire with a binary signal of either HIGH or LOW. The transceiver receives the binary signal that is on the SIGOUT wire.

NFC-WI is fully compliant and directly coupled with all modes, types, and data rates of ISO/IEC 18092 and ISO/IEC 14443, and no additional adaptation and no protocol conversion are required. It is a reliable concept that is feasible for immediate implementation.

> **NOTE** *For more information on NFC-WI, refer to ECMA International, ECMA 373: Near Field Communication Wired Interface (NFC-WI),* `www.ecma-international.org/memento/TC47-M.htm`.

SWP

The other physical interface option is SWP, which defines a single-wire connection between the SE and NFC controller in the mobile phone. Remember that NFC-WI uses a double-wire connection. ETSI TS 102 613 defines the SWP standard. SWP is a digital full-duplex protocol. The SWP interface is a bit-oriented and point-to-point communication protocol between an SE and NFC controller. The working principle is similar to that of master and slave: the NFC controller is comparable with the master, and the SE is comparable with the slave.

The SWP is mainly intended for use by UICC cards in mobile phones because only one of the standard eight contact paths is available for the SWP function. A special case occurs here, as shown in Figure 2-6, so that the voltage (Vcc) of the UICC card is not directly supplied by the mobile phone but is supplied through the NFC interface instead. This arrangement is necessary to enable contactless data transmission with SEs even when the battery is exhausted. If the NFC interface is close to an NFC reader, the reader field supplies energy to the SE through the NFC interface.

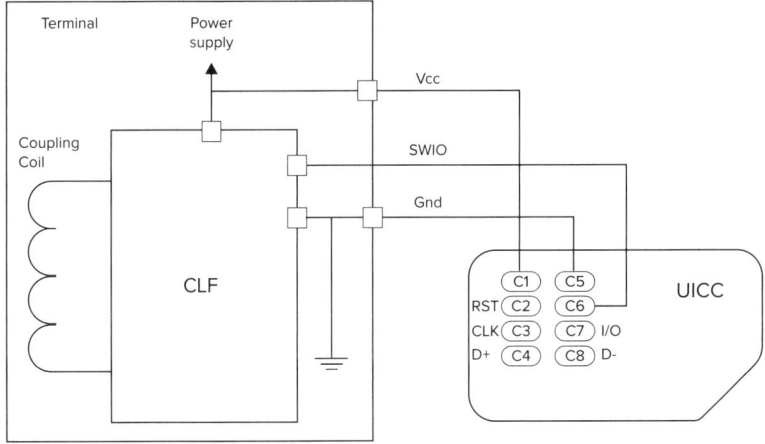

© European Telecommunications Standards Institute 2008. Further use, modification, copy and/or distribution are strictly prohibited. ETSI standards are available from http://pda.etsi.org/pda/.

FIGURE 2-6

> **NOTE** *For more information on SWP, refer to ETSI TS 102 613, Smart Cards; UICC — Contactless Front-end (CLF) Interface; Part 1: Physical and data link layer characteristics.*

HCI

HCI is a logical interface that allows an NFC interface to communicate directly with the application processor and SE. The HCI may be used in various electronic devices such as mobile devices, PDAs, and PC peripherals. For NFC-enabled mobile phones, it enables faster integration of NFC functionality. The HCI is standardized in ETSI TS 102 622.

The HCI defines the interface between logical entities called hosts that operate one or more services. According to the ETSI terminology, a network of two or more hosts is called a host network, and the host that is also responsible for managing a host network is called a host controller. In a host network that has a star topology, all hosts are connected to the host controller. The HCI has three components: a collection of gates that exchange commands, responses, and events; a Host Controller Protocol (HCP) messaging mechanism; and an HCP routing mechanism that may optionally segment messages when required. Figure 2-7 shows the simple HCP stack in a host network.

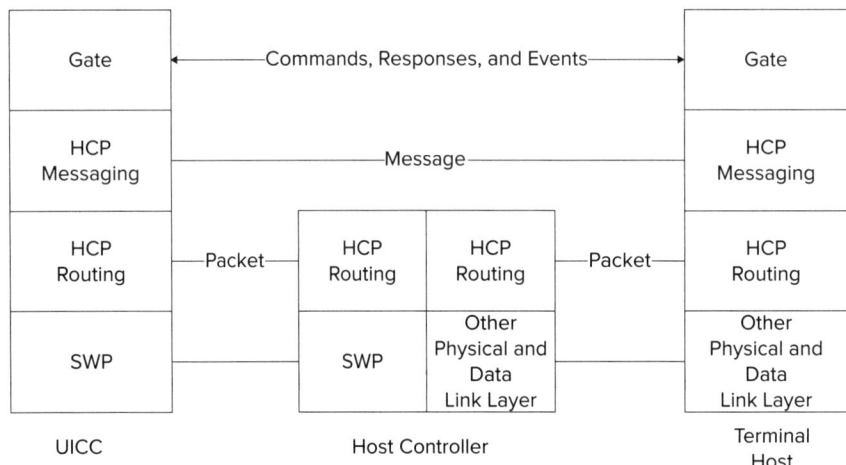

© European Telecommunications Standards Institute 2008. Further use, modification, copy and/or distribution are strictly prohibited. ETSI standards are available from http://pda.etsi.org/pda/.

FIGURE 2-7

> **NOTE** *For more information on HCI, refer to ETSI TS 102 622, Smart Cards; UICC — Contactless Front-end (CLF) Interface; Host Controller Interface (HCI), Technical Specification.*

STANDARDS USED BY NFC

As mentioned in Chapter 1, "Overview of Near Field Communication," NFC technology is based on RFID technology in a proximity range of 13.56 MHz. NFC enables data transfer up to 424 Kbps. The communication between two NFC devices is standardized as NFCIP-1 in the ISO/IEC 18092 standard. This standard defines only device-to-device communication for both active and passive communication modes.

The RF layer of NFC is a superset of the standard protocols, which are also compatible with the ISO/IEC 14443 standard (contactless proximity smart card standard) and JIS X 6319 standard as

FeliCa (another contactless proximity smart card standard by Sony) as well as the ISO/IEC 15693 standard (contactless vicinity smart card standard).

These smart card interfaces operate at 13.56 MHz from card reader to card with different data rates and communication ranges. Table 2-1 gives a short summary and comparison of ISO/IEC 14443, ISO/IEC 15693, and ISO/IEC 18092 communication interfaces. This section provides brief information about the proximity range communication interfaces and other important NFC standards.

TABLE 2-1: Summary of Communication Interface Standards

PARAMETERS	ISO/IEC 14443	ISO/IEC 15693	ISO/IEC 18092
Operating Mode	Reader to card	Reader to card	Peer-to-Peer
Communication Mode	Passive	Passive	Active and Passive
Range	Proximity	Vicinity	Proximity
Data Rate	106 Kbps	Up to 26 Kbps	106, 212, 424 Kbps

Proximity Contactless Smart Card Standards

As described in ISO/IEC 14443, proximity transactions are based on electromagnetic coupling between a proximity card and an RFID reader that uses an embedded microcontroller (including its own processor and one of several types of memory) and a magnetic loop antenna that operates at 13.56 MHz. ISO/IEC 14443 enables contactless transactions between a reader device and a proximity card used for identification.

The ISO/IEC 14443 standard contains four major parts (Table 2-2): the physical characteristics are explained in the first part of the standardization document, radio frequency power and signal interface are explained in the second part, initialization and anticollision protocols constitute the third part, and the transmission protocol is defined in the fourth part. This standard also defines two major contactless cards, namely Type A and Type B.

TABLE 2-2: Parts of ISO/IEC 14443 Standard

PART NAME	CONTENT
Part 1: Physical Characteristics of Contactless Smart Cards (PICC)	Defines physical characteristics of a contactless smart card, lists several requirements and tests that need to be performed at card level for construction of the card, antenna design, etc.
Part 2: RF Power and Signal Interface	Defines the RF power and signal interface, Type A and Type B signaling schemes, and also determines how the card is powered by RF field, etc.

continues

TABLE 2-2 *(continued)*

PART NAME	CONTENT
Part 3: Initialization and Anticollision	Defines the initialization and anticollision protocols for Type A and Type B, as well as anticollision commands, responses, data frame and timing.
Part 4: Transmission Protocol	Determines the high-level data transmission protocols for Type A and Type B, which are optional protocols, thus proximity cards may be designed with or without support for Part 4 protocols.

Major Proximity Contactless Smart Card Technologies

Up to now, various proximity coupling smart card technologies have emerged; however, only a few of them are compatible with the ISO/IEC 14443 standard. Currently, the most famous and competing contactless smart cards are MIFARE, Calypso, and FeliCa.

MIFARE

MIFARE is a well-known and widely used 13.56 MHz contactless proximity smart card system developed and owned by NXP Semiconductors, which is a spin-off company of Philips Semiconductors. MIFARE is an ISO/IEC 14443 Type A standard. The MIFARE family contains different types of cards, such as Ultralight, Standard, DESfire, Classic, Plus, and SmartMX. MIFARE Classic cards have varying memory sizes. MIFARE-based smart cards are being used in an increasingly broad range of applications, mostly in public transport ticketing and also for access management, e-payment, road tolling, and loyalty applications.

> **NOTE** *For more information on MIFARE products, visit the website* `www.MIFARE.net/products`.

FeliCa

FeliCa is a 13.56 MHz contactless proximity high-speed smart card system from Sony and is primarily used in electronic money cards. The name FeliCa comes from the word *felicity*, suggesting that the technology will make your daily life more convenient and enjoyable. FeliCa complies only with the Japanese Industrial Standard (JIS) X 6319 Part 4, which defines high-speed proximity cards.

> **NOTE** *For more information on FeliCa, visit the website* `www.sony.net/Products/felica/`.

Calypso

Calypso is another example of a contactless proximity smart card that conforms to the international public transportation standard. It was originally designed by a group of European transit operators from Belgium, Germany, France, Italy, and Portugal. It enables interoperability between several transport operators in the same area.

> **NOTE** *For more information on Calypso, visit the website* `www.calypsonet-asso.org/index.php.`

NFCIP

NFCIP (Near Field Communication Interface and Protocol) is standardized in two forms: NFCIP-1, which defines the NFC communication modes on the RF layer and other technical features of the RF layer, and NFCIP-2, which supports mode-switching by detecting and selecting one of the communication modes.

NFCIP-1

NFCIP-1 is standardized in ISO/IEC 18092, ECMA 340, and ETSI TS 102 190. This standard defines two communication modes as active and passive. It also defines the RF field, RF communication signal interface, general protocol flow, and initialization conditions for the supported data rates of 106, 212, and 424 Kbps in detail. Moreover, it defines the transport protocol, including protocol activation, data exchange protocol with frame architecture and error-detecting code calculation (CRC for both communication modes at each data rate), and protocol deactivation methods.

> **NOTE** *For more information on NFCIP-1, refer to ECMA 340: Near Field Communication Interface and Protocol (NFCIP-1),* `www.ecma-international` `.org/memento/TC47-M.htm.`

NFCIP-2

NFCIP-2 is a standard specified in ISO/IEC 21481, ECMA 352, and ETSI TS 102 312. The standard specifies the communication mode selection mechanism and is designed not to disturb any ongoing communication at 13.56 MHz for devices implementing ISO/IEC 18092 (NFCIP-1), ISO/IEC 14443 (such as MIFARE), or ISO/IEC 15693 (long-range vicinity communication by RFID tags).

> **NOTE** *For more information on NFCIP-2, refer to ECMA 352: Near Field Communication Interface and Protocol (NFCIP-2),* `www.ecma-international` `.org/memento/TC47-M.htm.`

NFC OPERATING MODE ESSENTIALS

In this section, technical essentials of three operating modes of NFC and their protocol stack architectures are described.

Reader/Writer Mode

In reader/writer operating mode, an active NFC-enabled mobile phone initiates the wireless communication and either reads or alters data stored in the NFC tags afterward. In reader operating mode, an NFC-enabled mobile phone is capable of reading NFC Forum–mandated tag types, such as NFC smart poster tags. This capability enables the mobile user to retrieve data stored in the tag and take appropriate actions afterward. The writer operating mode enables NFC mobiles to write data on NFC Forum–mandated tag types.

As illustrated in Figure 2-8, the reader/writer mode's RF interface is compliant with ISO/IEC 14443 Type A, Type B, and FeliCa schemes. The NFC Forum has standardized tag types, operation of those tag types, and the data exchange format between components. The reader/writer operating mode usually does not need a secure area — SE, in other words — of the NFC-enabled mobile phone. The process consists of only reading data stored inside the passive tag and writing data to the passive tag. The protocol stack architecture of the reader/writer operating mode, the NFC Data Exchange Format (NDEF), and record types are explained in the following sections.

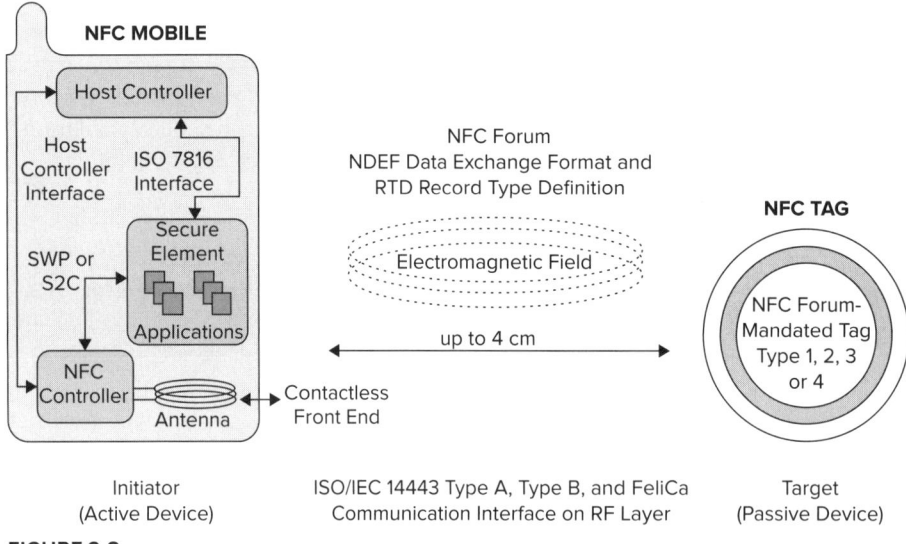

FIGURE 2-8

Protocol Stack Architecture

Figure 2-9 provides a useful protocol stack illustration of the reader/writer mode. The NFC device operating in reader/writer mode has the following protocol stack elements:

➤ Analog is related to RF characteristics of NFC devices and determines the operating range of devices. Digital protocols refer to the digital aspects of ISO/IEC 18092 and ISO/IEC 14443 standards and define building blocks of the communication. There is yet another important specification by the NFC Forum at this level: the NFC Activities Specification. This specification defines the required activities that set up communication in an interoperable manner based on a digital protocol specification such as polling cycles or when to perform collision detection.

➤ Tag operations indicate the commands and instructions NFC devices use to operate NFC Forum–mandated tags, which are Type 1, Type 2, Type 3, and Type 4. They enable read/write operations by using the NDEF data format and record type definitions (RTDs), such as smart poster, and URI RTDs from or to the tag.

➤ NDEF applications are based on NDEF specifications, such as using smart posters and reading product information from NFC-enabled smart-shopping fliers.

➤ Non-NDEF applications are vendor specific and are not based on NDEF standard specifications such as an electronic purse balance reader or contactless ticket reader.

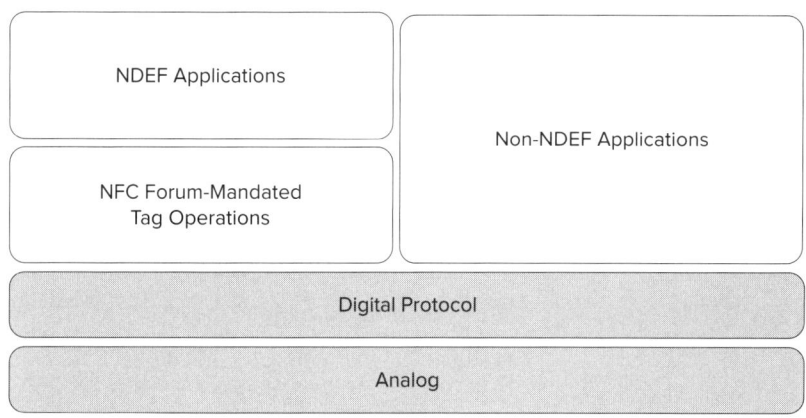

FIGURE 2-9

NFC Forum Tag Types

The NFC Forum defines four tag types and gives them designations between 1 and 4. Each tag type has a different format and capacity. NFC tag type formats are based on ISO/IEC 14443 Type A, ISO/IEC 14443 Type B, or Sony FeliCa. All currently available NFC tag type definitions are listed here, and Table 2-3 gives a summary of the tag types.

➤ **Type 1 tag:** The NFC Type 1 tag is based on the ISO/IEC 14443 Type A standard. These NFC tags are both readable and writable. Users can modify data on these tags and can configure the tags to become read-only when required. Memory availability is up to 1 KB, which is only just enough to store a website URL or similar amount of data. The memory size can be expanded up to 2 KB. The communication speed of this NFC tag is 106 Kbps. As a result of its simplicity, this tag type is cost effective and still can be used in most NFC applications.

➤ **Type 2 tag:** The NFC Type 2 tag also is based on the ISO/IEC 14443 Type A standard. These NFC tags are also both readable and writable, and users can configure them to become read-only when required. Again, the communication speed is 106 Kbps. The major difference between this tag and Type 1 tag is that its memory size is expanded up to 2 KB.

➤ **Type 3 tag:** The NFC Type 3 tag is based on the Sony FeliCa contactless smart card interface. It currently has a 2 KB memory capacity, and the data communications speed is 212 Kbps. This tag type is more suitable for complex applications, but it is more expensive compared with other tag types.

➤ **Type 4 tag**: The NFC Type 4 tag is compatible with both ISO 14443 Type A and Type B standards. These NFC tags are preconfigured during the manufacturing phase, and they are either writable or read-only; the type is defined during the manufacturing phase. The memory capacity can be up to 32 KB, and the communication speed is between 106 and 424 Kbps.

TABLE 2-3: Comparison of NFC Forum Tag Types

PARAMETER	TYPE 1	TYPE 2	TYPE 3	TYPE 4
Based On	ISO/IEC 14443 Type A	ISO/IEC 14443 Type A	FeliCa	ISO/IEC 14443 Type A, Type B
Chip Name	Topaz	MIFARE	FeliCa	DESFire, SmartMX-JCOP
Memory Size	Up to 1 KB	Up to 2 KB	Up to 1 MB	Up to 64 KB
Data Rate	106	106	212	106–424
Cost	Low	Low	High	Medium/High
Security	16- or 32-bit digital signature	Insecure	16- or 32-bit digital signature	Variable
Use Cases	Tags with small memory for a single application		Tags with small memory for a single application	

NDEF

NDEF is a data format to exchange information between NFC devices — namely, between an active NFC device and a passive tag or between two active NFC devices. The NDEF specification is a standard defined by the NFC Forum. The NDEF message is exchanged when an NFC device is in the proximity of an NFC Forum–mandated tag, the NDEF message is received from the NFC Forum–mandated tag, or it is received from the NFC Forum Logical Link Control Protocol (LLCP). The data format is the same in all cases.

NDEF is a binary message format that encapsulates one or more application-defined payloads into a single message. An NDEF message contains one or more NDEF records (see Figure 2-10). Each record consists of a payload up to $2^{32}-1$ octets in size. You can further chain together records to support larger payloads. Consequently, the maximum number of NDEF records that can be carried is unlimited.

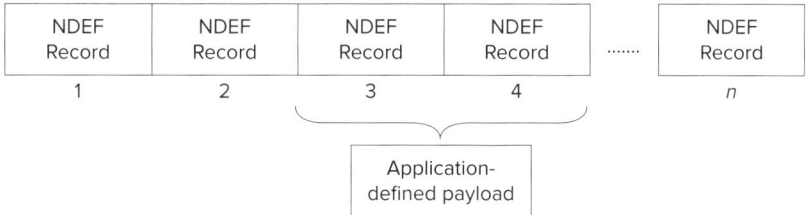

FIGURE 2-10

In an NDEF message, the first record is marked with the MB (Message Begin) flag set, and the last record is marked with the ME (Message End) flag set (see Figure 2-11). The minimum message length is one record, which you can achieve by setting both the MB and ME flags in the same record.

NDEF Message					
R_1 MB=1	...	R_r	...	R_s	R_t ME=1

FIGURE 2-11

The message head reads from left (head) to right (tail). Index 1 refers to the first flag set (MB), and index t refers to the last one (ME). So, the logical record indices are in the following order: $t > s > r > 1$ (as shown previously in Figure 2-11).

A record is the unit for carrying a payload within an NDEF message. Each NDEF record carries three parameters for describing its payload: payload length, payload type, and an optional payload identifier. The purposes of using these parameters are as follows:

➤ The payload length indicates the total number of octets in the current payload.

➤ The payload type identifier indicates the type of payload. NDEF supports URIs, MIME media type constructs, and an NFC-specific type format as type identifiers. By indicating the type of a payload, you are able to dispatch the payload to the appropriate application.

➤ The optional payload identifier allows user applications to identify the payload carried within an NDEF record.

NDEF records are variable length records with a common format, as illustrated in Figure 2-12. Each individual field in a record has different features. The details of each record are explained in the following list.

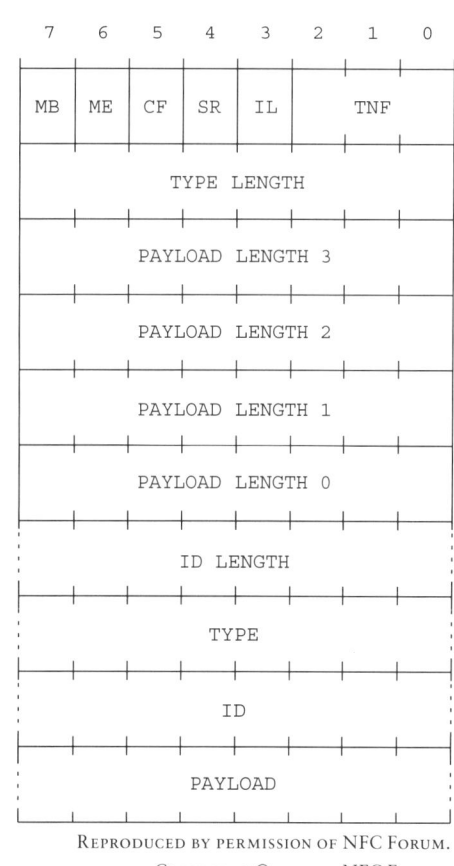

FIGURE 2-12

➤ **MB (Message Begin):** The MB flag is a 1-bit field which indicates the start of an NDEF message.

➤ **ME (Message End):** The ME flag is a 1-bit field which indicates the end of an NDEF message.

➤ **CF (Chunk Flag):** The CF flag is a 1-bit field which indicates that this is either the first-record chunk or a middle-record chunk of a chunked payload.

➤ **SR (Short Record):** The SR flag is a 1-bit field which indicates that the PAYLOAD_LENGTH field is a single octet.

➤ **IL:** The IL flag is a 1-bit field which indicates that the ID_LENGTH field is present in the header as a single octet.

➤ **TNF (Type Name Format):** The TNF field value represents the structure of the value of the TYPE field. The TNF field is a 3-bit field with values defined in Table 2-4.

➤ **TYPE_LENGTH:** This field is an unsigned 8-bit integer that specifies the length in octets of the TYPE field. The TYPE_LENGTH field is always zero for certain values of the TNF field.

➤ **ID_LENGTH:** This field is an unsigned 8-bit integer that specifies the length in octets of the ID field.

➤ **PAYLOAD_LENGTH:** This field is an unsigned integer that specifies the length in octets of the PAYLOAD field (the application payload). The size of the PAYLOAD_LENGTH field is determined by the value of the SR flag.

➤ **TYPE:** This field describes the type of the payload.

➤ **ID:** The value of the ID field is an identifier in the form of a URI reference.

➤ **PAYLOAD:** This field carries the payload intended for the NDEF user application.

TABLE 2-4: Type Name Format (TNF) Field Values

TYPE NAME FORMAT	VALUE
Empty	0x00
NFC Forum well-known type	0x01
Media-type	0x02
Absolute URI	0x03
NFC Forum external type	0x04
Unknown	0x05
Unchanged	0x06
Reserved	0x07

Record Types for NDEF

The NDF Forum defines various record types for NDEF messaging. The *record type* string field contains the name of the record type as *record type name*. NDEF applications use record type names to identify the semantics and structure of the record content. You can specify record type names in several formats in the TNF field of the NDEF record header. Record type names may be MIME media types, absolute URIs, NFC Forum external type names, or well-known NFC type names. Each record type definition is identified by its record type name:

➤ **NFC Forum Well-Known Type:** This dense format is designed for using tags and creating primitives or common types. You identify it inside an NDEF message by setting the TNF field of a record to the value 0×01. An NFC Forum well-known type is a URN (Uniform Resource Name) with the "nfc" namespace identifier (NID). The namespace-specific string (NSS) of an NFC Forum well-known type is prefixed with "wkt:" (see Table 2-5). When you encode it in an NDEF message, you must write it as a relative-URI construct in which the NID and "wkt:" prefixes are discarded. For instance, urn:nfc:wkt:very-complicated-type as a NFC Forum well-known type is encoded as very-complicated-type. The two major types are as follows:

➤ **NFC Forum Global Type:** The NFC Forum defines and manages this type. NFC Forum global types start with an uppercase letter in a character set that can be found in the RTD specification — for instance, A, Trip-to-Istanbul.

➤ **NFC Forum Local Type:** The NFC Forum local types start with a lowercase character in a character set or with a number in a character set that can be found in the RTD specification — for instance, 2, a, trip. An NFC Forum local type can be reused by another application in a different context and with a different content.

➤ **NFC Forum External Type:** The external type name is identified inside an NDEF message by setting the TNF field or a record to the value 0×04. The NFC Forum external type is a URN with the nfc NID. The NSS part is prefixed with ext: — for instance, urn:nfc:ext:yourcompany.com:f. As with NFC Forum well-known types, the binary encoding of NFC Forum external type discards the NID and NSS prefix ext:.

The NFC Forum defines various record types for specific cases: smart posters, URIs, digital signatures, and text. You can find more details on these RTDs on the NFC Forum's website. The following sections provide short descriptions for each one.

TABLE 2-5: Well-Known NDEF Record Type Examples

NDEF RECORD TYPE	DESCRIPTION	URI REFERENCE
Sp	Smart Poster	urn:nfc:wkt:Sp
T	Text	urn:nfc:wkt:T
U	URI	urn:nfc:wkt:U
Sig	Signature	urn:nfc:wkt:Sig

URI RTD

The URI Service RTD defines a record to be used with the NDEF to retrieve a URI stored on an NFC tag or to transport a URI from one NFC device to another. The well-known type for a URI record is "U" (0x55 in the NDEF binary representation). There are 256 URI identifier codes, the details of which you can find in the URI RTD specification of the NFC Forum. Some examples of URI identifier codes are shown in Table 2-6. All defined URIs are listed in Appendix A. The URI field is defined by the UTF-8 coding, which is also called NFC binary-encoding in specifications.

TABLE 2-6: Some Examples for URI Identifier Codes

DECIMAL	HEX	PROTOCOL TYPE
1	0x01	http://www.
2	0x02	https//www.
5	0x05	tel:
6	0x06	mailto:

Table 2-7 shows a simple example on launching NFC Lab-Istanbul's website, `http://www.nfclab.com`, where the URI identifier code field is `0x01`. If the content of this field is `0x02`, and the content of the URI field reads as `nfclab.com`, the resulting URI is `https://www.nfclab.com`.

TABLE 2-7: Storing a URL

OFFSET	CONTENT	EXPLANATION	FIELD
0	0xD1	SR=1, TNF=0x01 (NFC Forum well-known type), ME=1, MB=1	Header (MB, ME, CF, SR, IL, TNF)
1	0x01	Length of the Record Type (1 byte)	Type Length
2	0x0B	Length of the payload (11 bytes)	Payload Length
3	0x55	The URI record type ("U")	Type
4	0x01	URI identifier code as "http://www."	Payload
5	0x6e 0x66 0x63 0x6c 0x61 0x62 0x2e 0x63 0x6f 0x6d	The string "nfclab.com" in UTF-8	Payload

Smart Poster RTD

The Smart Poster RTD defines an NFC Forum well-known type on how to put URLs, SMSs, or phone numbers on an NFC Forum–mandated tag or how to transport them between devices. Smart posters are popular use cases for NFC-enabled applications. The idea is that an object can be made "smart" so that it is capable of storing additional information in the form of an NFC Forum–mandated tag. When a user touches an NFC device to the tag, this information can be read and processed afterward. The smart poster contains data that will trigger an application in the device such as launching a browser to view a website, sending an SMS to a premium service to receive a ring tone, and so on. The smart poster concept is mostly built around URIs, which became the standard for referencing information around the Internet. URIs are very powerful, and as already mentioned, they can represent anything from unique identifiers and web addresses to SMS messages, phone calls, and so on. The content of a smart poster payload is an NDEF message. The content of this message consists of several NDEF records. The most important ones are as follows:

➤ Title record for the service

➤ URI record that is the core of the smart poster

➤ Action record that describes how the service should be treated

➤ Icon record that refers to one or many MIME-type image records (icons) within the smart poster

➤ Size record for telling the size of the URI references that has an external entity such as a URL

➤ Type record for declaring the MIME type of the URI references that has an external entity such as a URL

As shown in Table 2-8, the content of this message represents a web address for the NFC Lab-Istanbul, so that when the user touches the tag on this smart poster, it triggers a browser on the NFC device and displays the NFC Lab-Istanbul's website.

TABLE 2-8: URI Example on a Smart Poster

OFFSET	CONTENT	LENGTH	EXPLANATION	FIELD
0	0xD1	1	TNF=0x01 (well-known type), SR=1, MB=1, ME=1	Header
1	0x02	1	Record Name Length (2 bytes)	Type Length
2	0x12	1	Length of the Smart Poster data (18 bytes)	Payload Length
3	Sp	2	The record name	Type
5	0xD1	1	TNF=0x01 (well-known type), SR=1, MB=1, ME=1	Header

continues

TABLE 2-8 *(continued)*

OFFSET	CONTENT	LENGTH	EXPLANATION	FIELD
6	0x01	1	Record Name Length (1 bytes)	Type Length
7	0x0A	1	Length of the URI payload (11 bytes)	Payload Length
8	U	1	Record Type: "U"	Type
9	0x01	1	Abbreviation: "http://www."	Payload
10	nfclab.com	10	The URI itself	Payload

Text RTD

The text record contains free-form plain text. The text record may appear as the sole record in an NDEF message, but in this case the behavior is undefined and the application should handle this occurrence. Typically, the text record should be used in conjunction with other record types to provide explanatory text. The NFC Forum well-known type for the text record is "T," which is 0x54 in UTF-8 encoding. In text record types, the text can be encoded in either UTF-8 or UTF-16, which is defined by the status byte in the text record. The text record is composed typically of an NDEF record header, a payload, and the actual body text in UTF format. In the payload, the status byte defines the encoding structure. Table 2-8 shows an example for storing a simple text record.

The status byte encoding is shown in Table 2-9. To discover the language code from the status byte, the status byte should be masked with the value 0x3F. The most significant bit defines the encoding scheme (UTF-8 or UTF-16), so that to decode the actual text, the most significant bit should be discovered from the status byte.

TABLE 2-9: Status Byte Encoding

BIT NUMBER	CONTENT
7	0: The text is encoded in UTF-8 1: The text is encoded in UTF-16
6	RFU (must be set to zero)
5...0	The length of the IANA language code

The text record is composed typically of an NDEF record header, a payload, and the actual body text in UTF format. In the payload, the status byte defines the encoding structure. Table 2-10 shows an example for storing a simple text record.

TABLE 2-10: Storing a Text Record

OFFSET	CONTENT	EXPLANATION	FIELD
0	Not Available	IL flag = 0 (no ID field), SF=1 (Short format)	NDEF Record Header
1	0x01	Record Name Length	
2	0x12	Payload Data Length (18 bytes)	
3	T	Binary encoding of the name	
4	0x02	Status byte, which means that this is UTF-8 and has a 2-byte language code	Payload
5	en	"en" is the ISO code for "English"	
7	Hello, world!	UTF-8 string	Text

> **NOTE** For more information on NDEF and RTDs, refer to NFC Forum specifications: NFC Data Exchange Format (NDEF), Technical Specification, Record Type Definition (RTD), Smart Poster Record Type Definition, Text Record Type Definition, URI Record Type Definition, Signature Record Type Definition, www.nfc-forum.org/specs/.

Peer-to-Peer Mode

In peer-to-peer mode, two NFC-enabled mobile phones establish a bidirectional connection to exchange information, as depicted in Figure 2-13. They can exchange virtual business cards, digital photos, or other kinds of data. ISO/IEC 18092 defines the standard for the peer-to-peer operating mode's RF communication interface as NFCIP-1.

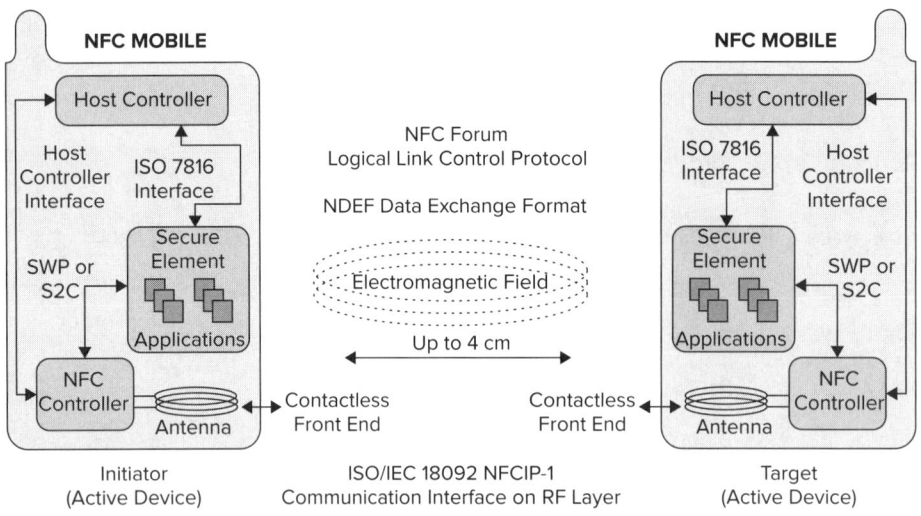

FIGURE 2-13

Protocol Stack Architecture

According to the NFC Forum specifications, an NFC device that is operating in peer-to-peer mode has the following protocol stack elements (see Figure 2-14):

➤ Analog and digital protocols are lower layer protocols standardized by NFCIP-1.

➤ LLCP allows transferring upper layer information units between two NFC devices.

➤ Protocol bindings provide standard bindings to NFC Forum protocols and allow interoperable use of registered protocols.

➤ NFC Forum protocols are the ones that the NFC Forum defines as binding to LLCP, such as OBEX and IP.

➤ The simple NDEF exchange protocol allows exchange of NDEF messages. It is also possible to run other protocols over the data link layer provided by LLCP.

➤ Applications may run over the simple NDEF exchange protocol, other protocols, or NFC Forum protocols. These applications include printing from a camera, exchanging business cards, and so on.

LLCP

The LLCP defines an OSI data link protocol to support peer-to-peer communication between two NFC-enabled devices. LLCP is essential for any NFC application that involves a bidirectional communication. LLCP provides a solid ground. It also enhances the basic functionalities provided by the NFCIP-1 protocol.

The NFCIP-1 protocol provides a Segmentation and Reassembly (SAR) capability, as well as data flow control, depending on the "Go and Wait" principle usual for half-duplex protocols. Furthermore, the NFCIP-1 protocol allows error handling by using an acknowledgment frame (ACK frame) and a negative acknowledgment frame (NACK frame) and provides an ordered data flow. It provides a link layer that is reliable and error free.

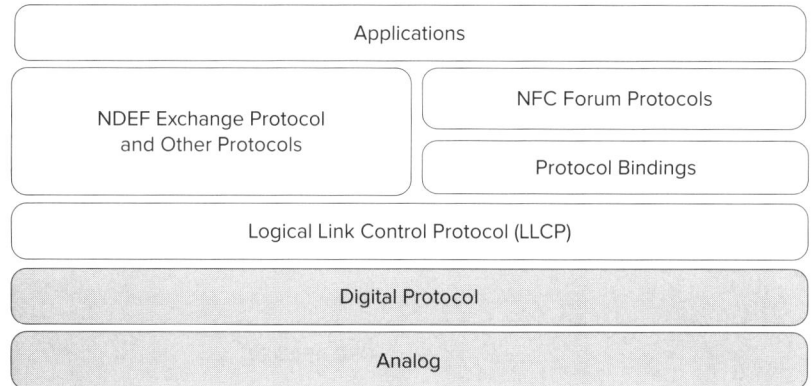

FIGURE 2-14

LLCP provides five important services: connectionless transport; connection-oriented transport; link activation, supervision, and deactivation; asynchronous balanced communication; and protocol multiplexing. They operate as follows:

➤ **Connectionless transport**: Connectionless transport provides an unacknowledged data transmission service. This transport mode can be used if upper protocol layers implement their own flow control mechanisms. Therefore, these layers do not need to rely on the data link layer's flow control mechanism.

➤ **Connection-oriented transport**: This mode provides a data transmission service with sequenced and guaranteed delivery of service data units. Data transmission is controlled through a sliding window protocol.

➤ **Link activation, supervision, and deactivation**: LLCP specifies how two NFC Forum devices within communication range recognize compatible LLCP implementations, establish an LLCP link, supervise the connection to the remote peer device, and deactivate the link when requested.

➤ **Asynchronous balanced communication**: Typical NFC MACs (Medium Access and Control) operate in Normal Response Mode where only a master, called the initiator, is allowed to send data to the target and also request data from the tag. The LLCP enables Asynchronous Balanced Mode (ABM) between service endpoints in the two peer devices by using a symmetry mechanism. Using ABM, service endpoints may initialize, supervise, recover from errors, and send information at any time.

➤ **Protocol multiplexing**: The LLCP is able to accommodate several instances of higher level protocols at the same time.

> **NOTE** *For more information on LLCP, refer to NFC Forum, Logical Link Control Protocol, Technical Specification, Version 1.1.*

Card Emulation Mode

In card emulation mode, the NFC-enabled mobile phone acts as a smart card. Either an NFC-enabled mobile phone that emulates an ISO/IEC 14443 smart card or a smart card chip integrated in a mobile phone is connected to the antenna of the NFC module. As the user touches a mobile phone to an NFC reader, the NFC reader initiates the communication immediately. The communication architecture of this mode is illustrated in Figure 2-15.

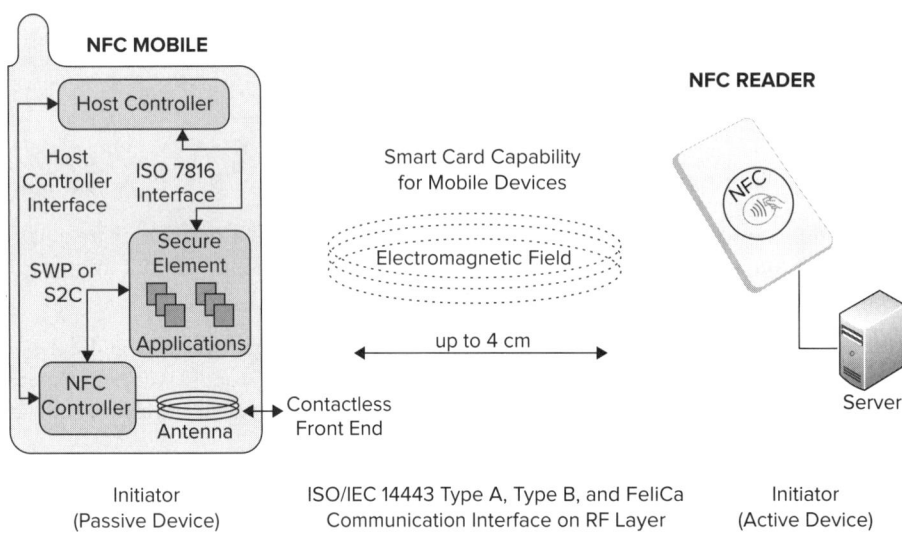

FIGURE 2-15

Protocol Stack Architecture

NFC devices that are operating in card emulation mode use similar digital protocol and analog techniques as smart cards, and they are completely compatible with the smart card standards (see Figure 2-16). Card emulation mode includes proprietary contactless card applications such as payment, ticketing, and access control. These applications are based on ISO/IEC 14443 Type A, Type B, and FeliCa communication interfaces.

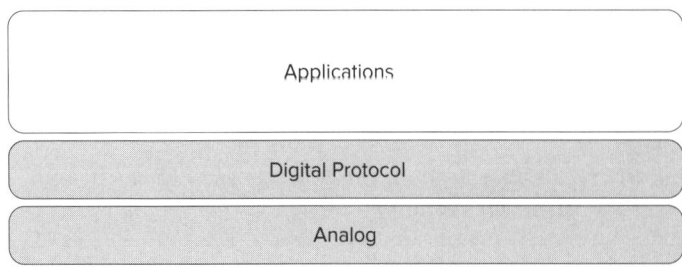

FIGURE 2-16

When an NFC reader interacts with an NFC mobile, the NFC mobile behaves like a standard smart card; thus, the NFC reader interacts with the payment applications on the SE. This scenario is depicted in Figure 2-15. Only card emulation mode uses an SE efficiently and securely, performing functions that require high security. Various studies and specifications, such as those by GlobalPlatform, GSMA, and Mobey Forum, are available to help you manage SE content remotely via over-the-air (OTA) technology.

STANDARDIZATION OF NFC

NFC technology benefits from various elements such as smart cards, mobile phones, card readers, and payment systems. All these elements need to acquire accreditation from an assortment of governing bodies that have the responsibility for the security and interoperability of various NFC devices. As mobile phones became the best solution for NFC technology, especially for secure transactions, various standardization bodies defined how the NFC technology should be integrated into mobile phones. Some other bodies defined the architectures and standards for the security as well as the ancillary technologies for NFC-enabled mobile phones, such as smart cards for NFC transactions. The common vision of all standardization bodies is to increase the ease of access, interoperability, and security for NFC technology. Table 2-11 gives a summary of the standardization bodies that play a role in the development of NFC technology and NFC mobile.

TABLE 2-11: Summary of Standardization Bodies

ORGANIZATION	STANDARDS OR ACTIVITIES GOVERNED WITHIN NFC ENVIRONMENT			RESPONSIBILITY
	ONLY MOBILE PHONES	SUPPORTIVE TECHNOLOGY	NFC TECHNOLOGY	
GSMA	✓	✓		Engages in technical, commercial, and public policy initiatives to ensure that mobile services are interoperable worldwide
OMA	✓	✓		Develops specifications for mobile service enabler to promote interoperability
JCP		✓	✓	Established specifications for the development of Java technology on mobile phones
ETSI and its Smart Card Platform	✓	✓		Develops globally applicable standards for information and communication technologies and handles SIM specifications

continues

TABLE 2-11 *(continued)*

ORGANIZATION	STANDARDS OR ACTIVITIES GOVERNED WITHIN NFC ENVIRONMENT			RESPONSIBILITY
	ONLY MOBILE PHONES	SUPPORTIVE TECHNOLOGY	NFC TECHNOLOGY	
3GPP	✓	✓		Develops globally applicable technical specifications for third-generation GSM
EMVCo		✓	✓	Provides specifications to ensure interoperability of smart card-based payment systems worldwide as well as mobile payment standards
Global Platform		✓	✓	Provides open and interoperable infrastructure and standards for transactions performed on smart cards
NFC Forum	✓	✓	✓	Develops specifications for NFC devices that are based on the ISO/IEC 18092 contactless interface ensuring interoperability among devices and services
ISO/IEC	✓	✓	✓	Provides worldwide international standards for business, government, and society
ECMA International	✓	✓	✓	Provides international standards and technical reports in order to facilitate and standardize the use of information communication technology and consumer electronics

Smart Card Alliance, "Security of Proximity Mobile Payments: A Smart Card Alliance Contactless and Mobile Payments Council White Paper," May 2009.

DIVERSITY OF NFC PLATFORMS

The NFC ecosystem includes a wide range of stakeholders or actors, depending on the types of provided services such as smart posters, payment, and ticketing. The type of NFC-enabled service defines the complexity in terms of which business model is applied, which stakeholders are involved, the appropriate collaboration model, the revenue model among players, and so on.

In this ecosystem, developing innovative NFC applications and deploying them for use by people are important issues that the participating actors within the ecosystem need to handle effectively.

To develop NFC applications, you need a complete understanding of NFC technology and operating modes. The two different types of applications in NFC services are graphical user interface (GUI) applications and SE applications. GUI applications exist in all operating mode applications and provide an interface that allows users to interact with a mobile device. This interface also provides the capability to read from and write to NFC components. In the case of SE applications, the application is needed to provide a secure and trusted environment for security-requiring applications such as payment, loyalty, and ticketing.

Today, various development tools on the market target different mobile phones. Some of these development tools include the Android software development kit (SDK) for Android mobile phones, Qt SDK for Symbian 3 mobile phones, Bada SDK for Bada operating system phones, and Series 40 Nokia 6212 NFC SDK for Nokia 6212 devices. Each development tool has a unique SDK and uses a different language. The developer who wants to develop an application on a specific platform needs to know that platform's programming language and NFC application programming interfaces (NFC APIs) built for that platform. Fortunately, today operating principles of different platforms' NFC APIs are similar to each other. Therefore, developers can easily work and develop NFC applications on different platforms.

In the case of deploying NFC applications, companies need to work together and create a unique, interoperable platform that will operate in all NFC mobiles and all SEs. Sustainable and collaborative ecosystem and business models are needed, and therefore, harmonizing the interests of all participants in these business models is essential. Without this, conflicting solutions that are not interoperable will be developed, and thus the technology cannot possibly be improved.

SUMMARY

NFC occurs between different types of smart NFC devices that can play either an initiator or a target role. These NFC devices are NFC-enabled mobile phones, NFC tags, and NFC readers. The NFC-enabled mobile phone is the major device for NFC transactions that exist in each interaction type. NFC-enabled mobile phones have a detailed, complex architecture with rich interfaces, including the NFC interface, which is the major component of NFC devices; secure elements with diverse alternatives; a host controller; and the host controller interface.

NFC operates at 13.56 MHz and transfers data up to 424 Kbps. Communication between two NFC devices is standardized in the ISO/IEC 18092 standard as NFCIP-1, which includes only device-to-device communication, peer-to-communication, and active/passive communication reader/writer modes. However, the RF layer of NFC communication is also compatible with the ISO/IEC 14443 standard (contactless proximity smart card standard), Japanese JIS X 6319 standard as FeliCa (another contactless proximity smart card standard by Sony), and ISO/IEC 15693 standard (contactless vicinity smart card standard). These smart card interfaces operate at 13.56 MHz with distinct data rates, communication ranges, as well as different modulation and coding features.

NFC technology offers three operating modes: reader/writer operating mode, peer-to-peer operating mode, and card emulation operating mode. Each operating mode's communications essentials are described in this chapter.

In reader/writer operating mode, an active NFC-enabled mobile phone initiates the wireless communication and reads and/or alters data stored in the NFC tag afterward. In this mode, an NFC-enabled mobile phone is capable of reading NFC Forum–mandated tag types. It is compatible with the ISO/IEC 14443 Type A, Type B, and FeliCa communication interface on the RF layer. The NFC Forum standardized the NFC Forum–mandated tag types as well as the NFC Data Exchange Format and various record types from the application layer to the RF layer.

In the case of peer-to-peer mode, communication occurs between two active NFC devices. One of the active devices initiates the communication, and a link-level communication is established between them thereafter. This mode uses the ISO/IEC 18092 NFCIP-1 standard for the communication. The NFC Forum standardized the Logical Link Control Protocol from the application layer to the physical layer.

Card emulation mode enables security- and privacy-required transactions with NFC. It gives smart card capability to mobile phones and uses ISO/IEC 14443 Type A, Type B, and FeliCa. The SE concept is an important issue in this mode to store and process valuable data and applications.

Because NFC technology relies on many other technologies, different components are needed to enable an NFC-based system, such as smart cards, NFC chips, tags, and development platforms for different vendors. For the development of NFC technology, various standardization bodies are providing specifications and standards to increase the ease of access, interoperability, and security of NFC technology and its supportive technologies. However, the standardization process is not yet complete. Therefore, various alternatives for different components of NFC systems have been produced. Thus, in the NFC ecosystem, developing innovative NFC applications and deploying them for use by people are important issues that the participating actors within the ecosystem need to handle effectively.

3

Getting Started with Android

WHAT'S IN THIS CHAPTER?

➤ Introduction to the Android operating system and its architecture

➤ Installation of Android SDK

➤ Structure and components of Android applications

➤ Intents and intent filters

➤ Fundamentals of the manifest file and application resources

➤ Essentials of the Dalvik Virtual Machine

➤ Background information for platform tools, SDK Tools, and Android Virtual Devices

This chapter provides introductory information to help you start development on Android platforms. It describes the preliminary requirements for Android application development, installation of the required development tools, and the Android application structure. This content is sufficient to help you start developing Android applications.

The information in this chapter also provides sufficient background information to help you understand the next chapter, "Android Software Development Primer." If you do not need information on the installation of the Android platform on computers, components of the Android operating system (OS), structure of Android applications, Android application packaging, Android SDK details, platform and SDK tools, or Android Virtual Devices (AVDs), you can skip this chapter and continue to Chapter 4.

The Android developers' official portal, `http://developer.android.com/`, is currently the resource most commonly used by Android developers in the world. We recommend you navigate through this site to get additional knowledge, resources, and up-to-date information on the Android platform as you need it.

Developing Android applications is mainly a platform-independent process. The programmers are free to use operating systems such as Windows, Linux, or Mac OS X. Accordingly, you may use any operating system you prefer. The software described in this book is not dependent on any OS and is usable on any one.

WHAT IS ANDROID?

Android is one of the world's most popular mobile platforms which is also used by mobile phones. Internally, it is a modified version of Linux with an integrated programming interface using the Java language. Android is owned by Open Handset Alliance, and it is indeed completely open sourced. Currently, Google leads the project, and most of the code is released under the Apache License. The application development environment includes a compiler, debugger, device emulator, and virtual machine named the Dalvik Virtual Machine (DVM) to execute the code.

If you are a Java programmer, you likely remember that Java codes are converted to byte codes and executed on a Java Virtual Machine (JVM) in the Java environment. Java byte codes, however, are not executed in Android. Instead, classes are complied into Dalvik executable files and run on the DVM. When you code your application, the Android SDK compiles the code into an Android package file with `.apk` extension which is to be uploaded and installed to the mobile device prior to execution.

Every Android mobile device has an installed Android OS version. The OS versions among different mobile devices may be different. When developing applications for specific Android devices, you should consider the platform version of that device. When you write an application for a targeted platform, some properties (for example, methods) may not be available for that specific version. Therefore, it is always a good practice to check the available resources in the application with the targeted platform's properties.

An Android application running on a mobile may perform all the services by itself or, alternatively, may use a previously available service on the mobile to perform some of the intended functionalities. For example, an application that requires a phone call may include the embedded code within the application or may fire — actually request firing — the already existing service on the mobile for this purpose.

As another example, if you implement an application that needs to record an image, it may make use of any available service that can take pictures. Consequently, you do not have to write any code for any service that is already available on the mobile. Your application simply activates the camera component; the component takes the picture and returns the result to your application. Activation of the camera service and data exchange is seamless, so the user does not notice that the main application uses a separate camera service. In other words, it is assumed that the camera service is a part of your application.

The major components of the Android OS are the Linux kernel, Android run time, libraries, application framework, and applications, as shown in Figure 3-1.

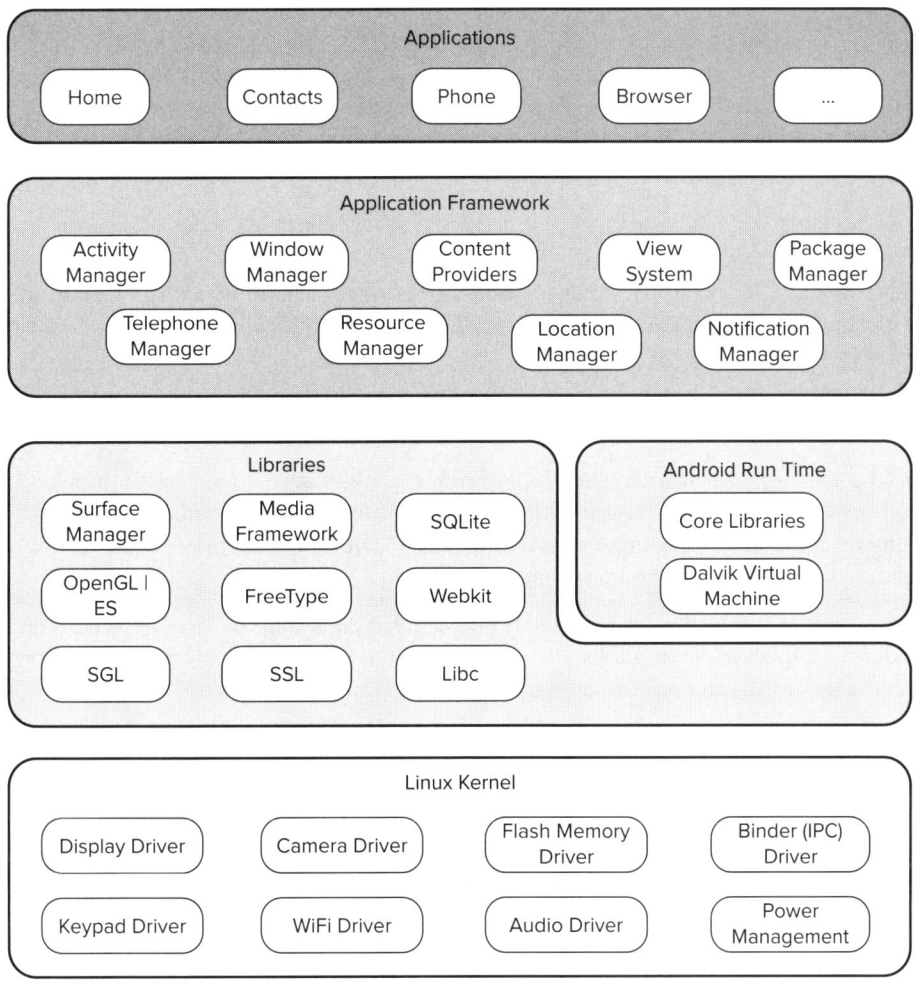

FIGURE 3-1

> **NOTE** *For more information on Android and its major components (Linux kernel, Android run time, libraries, application framework, and applications), refer to* `http://developer.android.com/guide/basics/what-is-android.html`.

Linux Kernel

The Linux kernel acts as an abstraction layer between the hardware and the software stack. Android consists of a Linux kernel version 2.6 to provide core system services. Starting with Android version 4.0 (Ice Cream Sandwich), Linux kernel version 3.x is being used.

Android Runtime

Android includes a set of core libraries that have most of the functionalities available in the core libraries of the Java programming language. Every Android application runs in its own process with its own instance of the DVM. Dalvik enables a device to run multiple VMs efficiently. DVM executes files in the Dalvik executable (`.dex`) format, which optimizes memory use.

Libraries

Android includes a set of C/C++ libraries that provide the necessary capabilities for the Android system. These capabilities are provided to developers through the application framework. Some of the core libraries are shown in Figure 3-1.

Application Framework

Because Android provides an open development platform, it enables developers to build rich and innovative applications. Developers can benefit from various Android components and services, such as accessing device hardware, using location information, running background services, setting alarms, and adding notifications to the status bar.

All the developers have access to the framework API, and the application architecture is designed in such a way that an application can publish its capabilities and then others can make use of those capabilities. Thus the reuse of components is simplified.

Applications

Android includes a set of core applications such as an e-mail client, SMS program, calendar, maps, browser, and contacts. All these applications are written using the Java programming language.

ANDROID SDK

For developers, the Android SDK provides a rich set of tools, including the debugger, libraries, handset emulator, documentation, sample code, and tutorials. Android applications can be easily developed using Eclipse (Android's official development platform). Eclipse provides rich features such as content assist, Java search, open resources, JUnit integration, and different views and perspectives for developing Android applications.

WHAT YOU NEED TO START

To develop Android applications, you need to install Java Development Kit (JDK) and Android Development Tools (ADT) Bundle.

Prior to installing the ADT Bundle to your computer, you should install JDK which also consists of the Java Runtime Environment (JRE).

ADT Bundle contains the required Android development tools and Eclipse as an Integrated Development Environment (IDE) which provides the necessary tools for computer programmers in order to develop new software.

Android also gives you the flexibility to integrate ADT into an existing IDE. At this time, you shouldn't download the bundle; instead you need to download ADT separately.

JDK and JRE

Android applications are developed in the Java programming language. Java is an object-oriented programming technology that is primarily designed to be object based and platform independent. It was introduced by Sun Microsystems in 1995 and is a trademark of the same company.

Java programs are written into files with the `.java` extension. Java programs are not converted to machine codes of the target platform for execution, but they are executed on interpreters instead. The Java interpreter varies according to the execution platforms. The converters on personal computers and similar machines are named Java Virtual Machines (JVMs). Before running a Java program with `.java` extension, you should compile and convert it to byte code with the `.class` extension using a Java compiler. Java byte code programs are then executed on top of the virtual machine.

Java technology is made up of the following three elements, the combination of which composes the Java platform:

➤ Java programming language *standards*, which define the rules to write a Java program

➤ *JVM* to execute Java programs written by satisfying the Java language standards

➤ An extensive set of *application programming interfaces (APIs)* that support a wide range of sources that a programmer needs

To install the JDK package, download the related package from `http://www.oracle.com/technetwork/java/javase/downloads/` and install it on your computer.

Android SDK

After installation of the Java SDK, you may install the Android SDK. Android SDK can be freely downloaded from `http://developer.android.com/sdk/`. From this site, you may download the ADT Bundle for any OS or download the SDK tools if you already have an existing IDE.

If you download the ADT Bundle, you will have all the related tools configured, so we recommend that you install the SDK using this bundle.

When you download the ADT Bundle and extract it, you will have two folders named `Eclipse` and `sdk`. The `Eclipse` folder contains the Eclipse IDE that you will use for software development. The `sdk` folder contains the latest Android tools and latest Android platform.

In order to run Eclipse, you simply click on `Eclipse.exe` in the `Eclipse` folder. When you run Eclipse, it needs to know where to write and read the project files, so it asks you to enter a *workspace* folder to store your Eclipse projects and all related development files. You may create a workspace folder anywhere on your computer or on the network where your computer is connected. To enter a name to be used as a default folder for all projects, you should select it as the default workspace (see Figure 3-2). In this case that folder is used for all further projects without asking you to enter another for each project. After you select a workspace, you see the Android IDE welcome screen that informs you of the successful installation (see Figure 3-3).

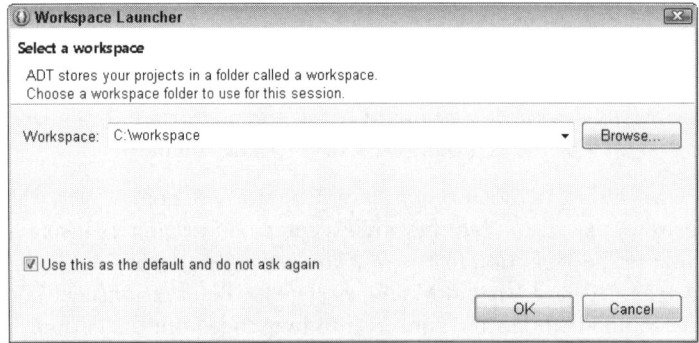

FIGURE 3-2

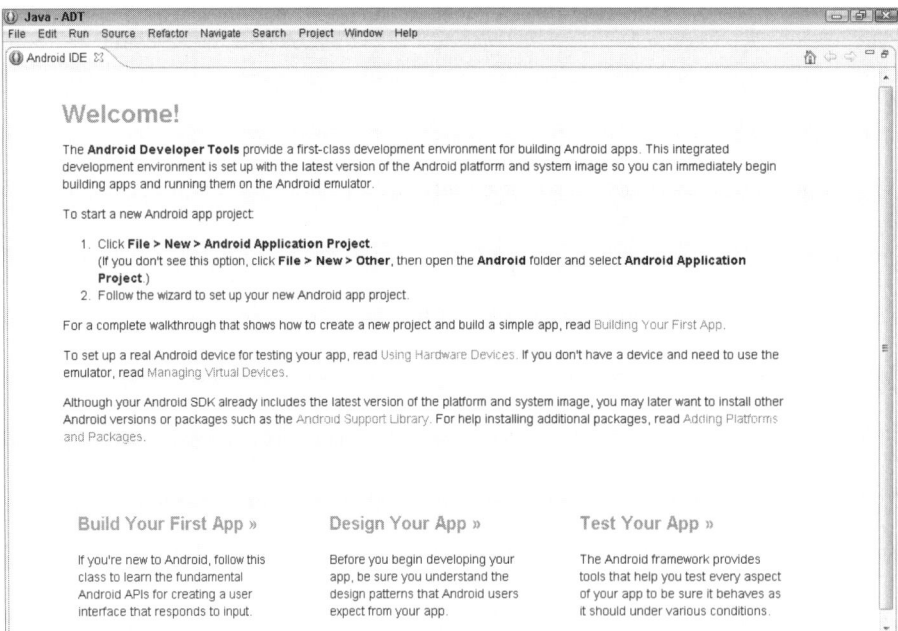

FIGURE 3-3

> **NOTE** *If you download SDK tools separately for an existing IDE, you need to perform additional steps. Please refer to the steps described at* `http://devel-oper.android.com/sdk/installing/`.

Adding More Platforms and Other Components to the SDK

Android SDK initially contains only the latest Android platform and required tools. However, you may also need to develop and test applications for an earlier Android SDK platform that matches your target device. Therefore, the last step in setting up your SDK is installing these components.

To install components, you should run the SDK Manager using one of the following methods:

➤ From ADT (Eclipse), select Window ➪ Android SDK Manager.

➤ On Windows, double-click the SDK Manager.exe file.

➤ On Mac OS X or Linux, open a terminal and navigate to the sdk/tools/ directory in the Android SDK; then execute android.

To install or update the components, you should click the checkbox next to them and click Install (see Figure 3-4). There is not any inconvenience to install more than one Android platform. Especially if you consider deploying your application to different platforms, it is better to install all the platforms.

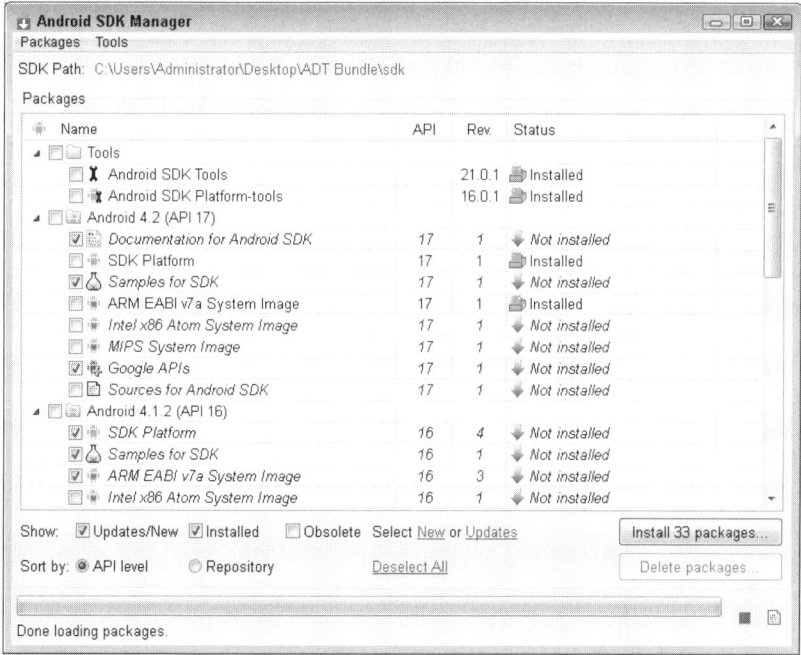

FIGURE 3-4

You can check the success of the installation from the SDK Manager (see Figure 3-5). You also can see installed SDKs from ADT (Eclipse) by opening Window ➪ Preferences (Mac OS X: Eclipse ➪ Preferences) and by selecting Android from the left panel (see Figure 3-6). You can identify the installed components from the Target Name header for this purpose.

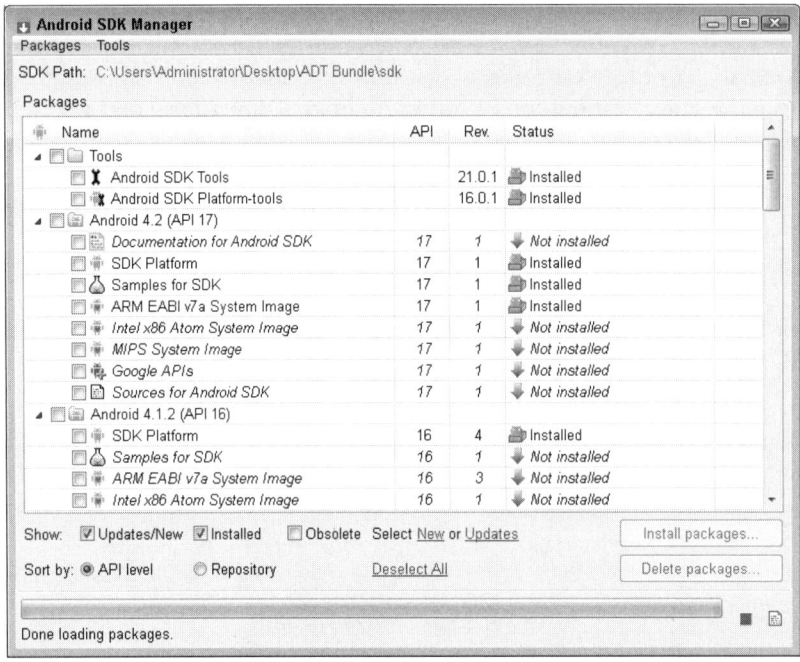

FIGURE 3-5

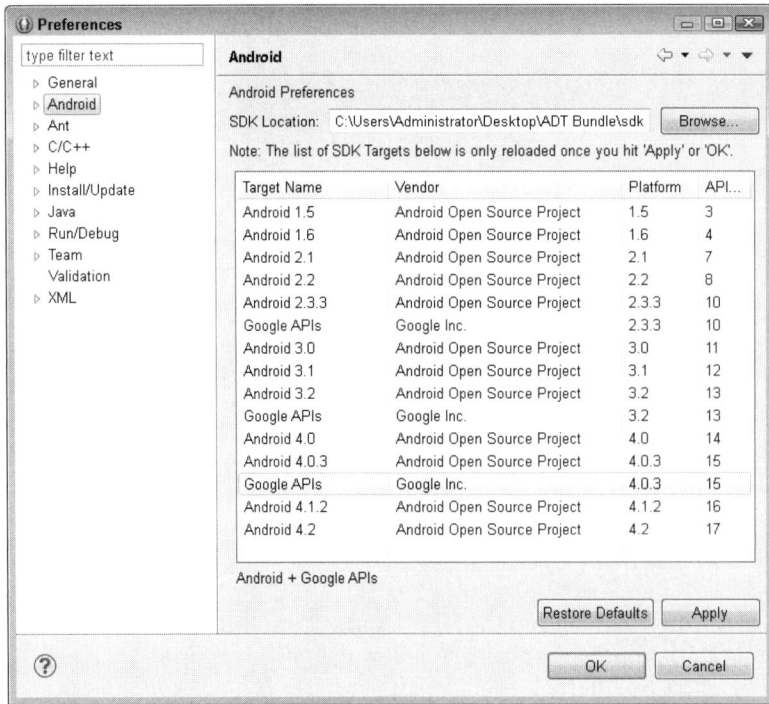

FIGURE 3-6

When you launch Android SDK Manager, you will see that various components exist. Most popular components are described in Table 3-1.

TABLE 3-1: Android SDK Package Components

SDK COMPONENT	DEFINITION	REQUIRED
SDK Tools	This package contains the required tools to debug and test your applications. These tools are located in the `tools` directory under the root directory of Android SDK (`sdk/tools/`).	SDK Tools is a required package. It is installed as part of the downloaded default SDK package. It should be updated as a new update is provided.
SDK Platform-tools	These tools are platform dependent and are needed to develop Android applications. These tools are located in the `platform-tools` directory under the root directory of Android SDK (`sdk/platform-tools/`).	SDK Platform-tools is another required package. It is installed as part of the downloaded default SDK package.
Android Platforms	An SDK platform actually corresponds to an Android platform. Each platform includes a system image, libraries, and sample codes. An Android Virtual Device (AVD) is also created based on an SDK platform. Platforms are located in the `platforms` directory under the root directory of Android SDK (`sdk/platforms/`).	At least one Android platform must be installed into your environment. It is reasonable to download and install the latest platform version. After developing the application, you also should install older platforms and test under those platforms if you plan to publish the application for older versions.
Samples	This package contains the sample applications with their source code. They are available for each platform. Samples are located in the `platforms` directory under the root directory of Android SDK (`sdk/platforms/platform-X/`).	The Samples package is not mandatory; however, it is nice to have one. You can load an example as a project and then run and test it. You may also use some parts of it in your own application as required.
Documentation	This package contains the documentation for the SDK. After downloading the package, you can work with the documentation offline. Documentation is located in the `platforms` directory under the root directory of Android SDK (`sdk/platforms/platform-X/`).	The Documentation package is not mandatory, but it is a useful one because it enables you to work offline with the API reference.

continues

TABLE 3-1 *(continued)*

SDK COMPONENT	DEFINITION	REQUIRED
Google APIs	This package contains the Google Maps external library API.	The Google APIs package enables your application to access the Maps external library. You can display and manipulate Maps data in your application.
USB Driver for Windows	This package contains the driver files to run and debug your application in a real Android device. Under Linux and Mac OS X, you do not need this driver.	You need the USB Driver package if you are developing on Windows and want to run and debug your application in a corresponding Android device.

SDK PACKAGES

After you have successfully downloaded and extracted ADT Bundle and installed additional components, many files are installed in several directories under `sdk` directory. Table 3-2 gives the description and the contents of each directory.

TABLE 3-2: Directories

FOLDER	DESCRIPTION
`add-ons/`	Contains the add-ons to the Android SDK
`docs/`	Contains a full set of documentation including the Developer's Guide and the API Reference
`platform-tools/`	Contains Android SDK Platform-tools
`platforms/`	Contains the installed Android platforms; each platform is displayed in a separate directory
`samples/`	Contains the installed SDK package for Samples; each platform's Samples are displayed in a separate directory
`tools/`	Contains the default SDK package for Tools

Inside the `tools` and `platform-tools` directories, you can find executable files for various objectives. To run these executable files from the command line without providing a full path name, you should add these folders to the `PATH` environment of the operating system.

Depending on your installed operating system, you can include these directories in your PATH in several ways:

➤ **On Windows:** Click Start and then right click My Computer and then select Properties. Click Advanced System Settings ➪ Environment Variables. Find the Path under System Variables and double-click it. At the end of the line, insert a semicolon, add the path to the `tools` directory, add a semicolon, and then add the path to the `platform-tools` directory as follows:

    ….;C:\android-sdk\tools;C:\android-sdk\platform-tools

➤ **On Linux:** Edit either the `.bash_profile` or `.bashrc` file in your home directory. Locate the line that sets the PATH environment variable and add the `tools` and `platform-tools` directories. If you do not find one, you can add one as follows:

    export PATH=$PATH:~/android-sdk/tools:~/android-sdk/ /platform-tools

➤ **On Mac OS X:** Look in your home directory for the `.bash_profile` file and proceed as described for Linux. You can create the `.bash_profile` file if you do not have one.

ANDROID API LEVELS

The Android OS's platform versions are identified by API level numbers. It is important to know the API level of the target Android OS device before attempting to write an application for it. When you develop an application for a specific targeted mobile device, you should check the Android OS's API level of the target mobile device to ensure the project will work after production. An API level is an integer greater than zero that uniquely identifies the corresponding Android platform.

Android applications are upward compatible, which ensures that when an application runs in an Android API level, say X, it also runs on all Android devices with higher API levels of X. Applications developed in a lower API level are compatible with higher level platforms. There is one exception to this, though: If a component available on that API level X is removed from the API level Y where Y > X, the application, or that component, depending on the case, does not run on the devices with API levels Y and higher. On the other hand, downward compatibility is not provided. It is important to note that the user of the actual mobile device may upgrade the mobile phone to newer versions of the Android OS for some reason, but downgrading the device to a lower version is unlikely. As a developer, you should take into account this possibility and use only components that are still available, are targeted, and use higher API levels, but should not use deprecated components.

Table 3-3 represents currently available API levels. Remember that at the installation stage of the SDK, API levels are displayed on the screen (as previously shown in Figure 3-4).

TABLE 3-3: Android API Levels

PLATFORM VERSION	API LEVEL	VERSION_CODE
Android 4.2	17	JELLY_BEAN_MR1
Android 4.1, 4.1.1	16	JELLY_BEAN
Android 4.0.3 , 4.0.4	15	ICE_CREAM_SANDWICH_MR1
Android 4.0, 4.0.1, 4.0.2	14	ICE_CREAM_SANDWICH
Android 3.2	13	HONEYCOMB_MR2
Android 3.1.x	12	HONEYCOMB_MR1
Android 3.0.x	11	HONEYCOMB
Android 2.3.4, 2.3.3	10	GINGERBREAD_MR1
Android 2.3.2, 2.3.1, 2.3	9	GINGERBREAD
Android 2.2.x	8	FROYO
Android 2.1.x	7	ECLAIR_MR1
Android 2.0.1	6	ECLAIR_0_1
Android 2.0	5	ECLAIR
Android 1.6	4	DONUT
Android 1.5	3	CUPCAKE
Android 1.1	2	BASE_1_1
Android 1.0	1	BASE

Android packages are indexed on the Android reference site: `http://developer.android.com/reference/`. When you navigate through packages, classes, and methods at the site, on the right side of the header, you can see their availability in terms of API levels (see Figure 3-7). For example, the `android` package and `android.Manifest.permission` classes have been available since API level 1; however, the `WRITE_EXTERNAL_STORAGE` constant inside the `android.Manifest.permission` class has been available only since API level 4.

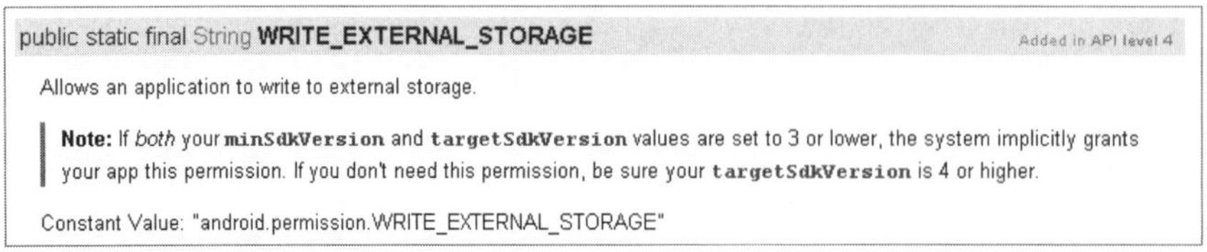
public static final String **WRITE_EXTERNAL_STORAGE** Added in API level 4

Allows an application to write to external storage.

Note: If *both* your **minSdkVersion** and **targetSdkVersion** values are set to 3 or lower, the system implicitly grants your app this permission. If you don't need this permission, be sure your **targetSdkVersion** is 4 or higher.

Constant Value: "android.permission.WRITE_EXTERNAL_STORAGE"

FIGURE 3-7

> **NOTE** *For more information on Android API levels, refer to* `http://developer`
> `.android.com/guide/appendix/api-levels.html.`

STRUCTURE OF ANDROID APPLICATIONS

Android applications are formed from various components. Understanding these components is an important step toward developing applications. You need and use these components at the development stage. The following sections describe these components and their functions.

Android Application Components

The four different building blocks of Android applications are *activities, services, broadcast receivers,* and *content providers.* Each Android application uses at least one activity and also may further use any number of other components. Furthermore, *intents and intent filters,* a *manifest file,* and *applications resources* are used along with these components.

Activities

Activities represent the user interface of the application. An application contains at least one activity and may use additional activities as required. Each activity presents one screen view to the user; therefore, the number of activities is equal to the number of different views required. With triggering actions, different screens are displayed to the user. For example, if you consider a phonebook application, the list of the addresses in the book is one activity. If the user selects one of the contacts from the initial view that is provided by one activity, the contact details may be displayed on another screen, which indeed is another activity.

In an Android application, typically there should be at least one main activity, which is the initial screen that appears when the user launches the application. The application obviously may trigger other activities as required.

Services

A service is a background task that does not provide any user interface to the user but performs operations in the background. Services persist even if the user exits from the application. For example, a mail service may continue getting e-mails when you play a game, or a GPS application may continue to fix the mobile's position when a phone call is made.

Android services can take two forms, so a service can be in one form of it at any given time:

➤ **Started:** A service is *started* when an application component invokes it using the `startService()` method. The service keeps running in the background, even if the component that invoked the service is killed. For a service to stop, the service itself should complete the intended operation. A service in the *started* form does not return a result to the caller component. A network operation may be an example of this kind of service.

➤ **Bound:** A service enters into a *bound* form when an application component calls it using the `bindService()` method. This form of service interacts with application components such as returning results to the application and sending requests. Multiple components can bind to this type of service, and when all the components unbind the service, the bound service is destroyed.

Broadcast Receivers

Broadcast receivers are the components that receive and react to broadcast announcements. Applications receive those broadcasts as they are announced, and sometimes applications need to react accordingly. Most of the broadcasts originate from the operating system, such as a time zone change, low battery announcement, or turned-off screen. Broadcasts do not have user interfaces; however, they may create a notification in the Android status bar.

Content Providers

A content provider makes an application's data available to other platforms as required. In another way, it can store and retrieve data, and make it available to other applications. Using content providers, your application can share data with others, and other applications may even modify it if your application allows such changes. Android contains the SQLite database, which can be used in this context.

Intents

Activities, services, and broadcast receivers are triggered by intents. For activities and services, intents define the actions to perform. For example, a service may make a request to open a webpage. For broadcast receivers, an intent defines the announcement that will be broadcast. However, content providers cannot be activated by intents.

Intents can be either explicit or implicit. An explicit intent calls a specific service or activity explicitly; whereas an implicit intent just gives the definition of the required service, and the Android OS selects one appropriate service among the registered available services. To register an intent of an application, you use an intent filter.

Intent Filters

Activities, services, and broadcast receivers can have one or more intent filters to inform the system. Each intent filter describes a capability of the component and a set of intents that the component is willing to receive. An intent filter filters the desired implicit intents, and unwanted implicit intents are filtered out automatically. Only the implicit intents that pass from the intent filters are delivered to the component. However, explicit intents are not filtered by the intent filters.

An intent filter is an instance of the `IntentFilter` class. Intent filters are defined in the application's manifest file inside `<intent-filter>` elements. The action field of the intent filter is mandatory, whereas data and category fields are optional. If all the fields are defined, the intents are tested for all filters and must pass all filters to be delivered to the component.

Manifest File

Before the Android OS starts an application component, the system must know that the application component exists by reading the application's manifest file, `AndroidManifest.xml`. This file includes information about all components of the corresponding application, which is at the root of the application project directory. In addition to declaration of application components such as activities, the manifest file also provides various other functions:

➤ Identifying user permissions that the application requires, such as Internet access

➤ Declaring minimum API levels required by the application based on APIs used by the application

➤ Declaring hardware and software features used or required by the application, such as camera and Bluetooth services

➤ Providing API libraries that the application needs to be linked, such as the Google Maps library

> **NOTE** *For more information on application fundamentals and the manifest file, refer to* `http://developer.android.com/guide/topics/fundamentals.html`.

The primary task of the manifest file is to inform the system about the application's components. Following is a manifest file example for declaring an activity's components:

```xml
<?xml version="1.0" encoding="utf-8"?>
<manifest ... >
   <application android:icon="@drawable/app_icon.png" ... >
      <activity android:name="com.example.project.ExampleActivity"
         android:label="@string/example_label" ... >
      </activity>
      ...
   </application>
</manifest>
```

According to the example, the `android:icon` attribute in the `<application>` element refers to the resources for an icon that identifies the application. In the `<activity>` element, the `android:name` attribute specifies the fully qualified class name of an activity. The `android:label` attribute specifies a string to use as the user-visible label for the activity. All application components need to be declared in the manifest file through the following elements:

➤ `<activity>`: elements for activities

➤ `<service>`: elements for services

➤ `<receiver>`: elements for broadcast receivers

➤ `<provider>`: elements for content providers

> **NOTE** *For more information about how to structure the manifest file for your application, refer to* http://developer.android.com/guide/topics/manifest/manifest-intro.html.

Another important note is that in activating components, you can use intents to start activities, services, and broadcast receivers. To declare a component in your application's manifest, you can optionally include intent filters that declare the capabilities of the component. An intent filter can be declared for a component through the `<intent-filter>` element as a child of the component's declaration element.

> **NOTE** *For more information about declaring intent filters in the manifest file, refer to* http://developer.android.com/guide/topics/intents/intents-filters.html.

Application Requirements

Currently, various devices powered by Android have different features and capabilities. To prevent your applications from being installed on the devices that your application cannot run, you should define the device and software requirements in a manifest file. For example, if your application requires NFC service and uses APIs introduced in Android 2.3.4, you need to declare them as requirements in the manifest file of the application. So, devices that do not have NFC capability and have an Android version lower than 2.3.4 cannot install your application from Google Play (previously known as Android Market). Google Play uses these declarations to filter which applications can be installed to the user's device.

Screen Size and Density

You need to consider two important device characteristics while designing and developing an Android application: screen size and screen density.

Screen size defines the physical dimensions of the screen, and screen density defines the physical density of the pixels on the screen or dots per inch. To simplify all the different types of screen configurations, the Android system generalizes them into categorical groups. The screen size options are small, normal, large, and extra large; and the screen density options are low density, medium density, high density, and extra high density. Although, by default, your application is compatible with all screen sizes and densities, you need to create customized layouts for certain screen sizes and provide images for certain densities to increase efficiency. Therefore, you need to declare these in the manifest file of your application through the `<supports-screens>` element.

Input Configurations

Devices provide various input mechanisms such as a hardware keyboard, trackball, and five-way navigation pad. If your application requires input hardware, you need to declare it in the manifest file of the application using the `<uses-configuration>` element.

Device Features

Many hardware and software features such as a camera, light sensor, Bluetooth, or touch screen may exist in an Android device. In this case, you need to use the `<uses-feature>` element for declaring device features of your application in your manifest file.

Platform Version

The relationships between Android OS versions and Android API levels were defined previously. Upward compatibility of Android applications was also explained. Be aware that you need to declare a minimum API level using the `<uses-sdk>` element. This API level needs to be the minimum Android API level that your application can run on.

Application Resources

An Android application is composed of various resources that are separate from the source code, such as images, audio files, and anything relating to the visual presentation of the application. For an Android application, you need to define animations, menus, styles, colors, and the layout of user interfaces within the related XML files.

Using application resources helps you update various characteristics of your application easily without modifying your source code. These resources enable you to optimize your application for a variety of device configurations such as different languages and screen sizes. For every resource that is included in an Android application, the SDK build tools define a unique integer ID. This ID can be used to reference the resource from your application code or from other resources defined in XML.

> **NOTE** *For more details about different kinds of resources that you can include in your Android application and how to create alternative resources for various device configurations, refer to* `http://developer.android.com/guide/topics/resources/index.html`.

Processes and Threads

When an application component is invoked and if the application does not have any other components running, the Android system starts a new process for the application with a single thread of execution. By default, all components of the same application run in the same process and thread (called the "main" thread). If an application component starts and a process already exists for that application (because another component from the application exists), the component is started within that process and uses the same thread of execution. However, you can arrange for different components in your application to run in separate processes, and you can create additional threads for any process.

The Android system tries to maintain an application process and needs to remove old processes to reclaim memory for new or more important processes. When a memory chunk is required for a new process, to determine which processes to keep and which processes to kill, the system assesses the *importance hierarchy* for each process based on the components running in the process and the state of those components. Processes with the lowest importance are discarded first, and then those with the next lowest importance are discarded, and so on.

Processes are divided into five levels of importance. Each process type is described here according to its order of importance.

Foreground Process

The foreground process is directly related to what the user is currently doing. Generally, only a few foreground processes exist at any given time. When the device reaches a memory paging state, some of the foreground processes need to be killed to keep the user interface responsive.

A process is considered to be a foreground process if it hosts any of the following activities or services:

➤ An activity that the user is currently interacting with, where the activity has called the `onResume()` method

➤ A service that is bound to the activity that the user is currently interacting with

➤ A service that is running in the foreground where the service has called `startForeground()`

➤ A service that is executing one of its life-cycle callbacks: `onCreate()`, `onStart()`, or `onDestroy()`

➤ A broadcast receiver that's executing its `onReceive()` method

Visible Process

A visible process does not have any foreground components, but it still can affect what the user sees on the screen. Thus, a visible process is considered as the next most important after foreground processes.

A process is considered to be visible if it hosts any activity or service presented here:

➤ An activity that is not in the foreground but is still visible to the user where the `onPause()` method has been called

➤ A service that is bound to a visible (or foreground) activity

Service Process

A service process is a service that has been started with the `startService()` method. The system keeps service processes running until there is not enough memory to retain them along with all foreground and visible processes.

Background Process

A background process runs an activity that is not visible to the user where the activity has called the `onStop()` method. These processes have no direct impact on the user experience, and the system can kill them any time to reclaim memory for a foreground, visible, or service process.

Empty Process

An empty process does not hold any active application components. The only reason to keep this kind of process alive is for caching purposes, to improve startup time the next time the same component needs to run. The system often kills these processes to balance overall system resources between process caches and the underlying kernel caches.

> **NOTE** *For more information on processes and threads, refer to* http://developer.android.com/guide/topics/fundamentals/processes-and-threads.html.

DALVIK VIRTUAL MACHINE (DVM)

As mentioned previously, Android uses a special virtual machine named DVM. This virtual machine uses a special byte code system called *dex bytecode* developed specifically for Android executable files.

When an Android application is implemented, the Java compiler converts the code into bytecode (`class` file), and then a "dx" converter converts it into Dalvik executables (`dex`). Finally, the `dex` file is packed into an Android package file (`apk` file) by using an archiver prior to installing it on a mobile phone (see Figure 3-8). When the Eclipse IDE and ADT plug-ins are used, the process to convert the code to an `apk` file is automated (see Figure 3-9). As a matter of fact, an `apk` file is a renamed `zip` file, so you may rename its extension to `zip` and then open to see its contents using a decompression utility.

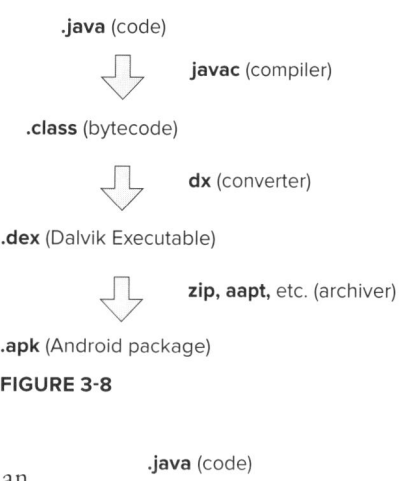

.java (code)

javac (compiler)

.class (bytecode)

dx (converter)

.dex (Dalvik Executable)

zip, aapt, etc. (archiver)

.apk (Android package)

FIGURE 3-8

.java (code)

Android IDE

.apk (Android Package)

FIGURE 3-9

PLATFORM TOOLS

Platform tools are typically updated every time you install a new SDK platform. Each update of the platform tools is backward compatible with older platforms. Platform tools are located in the `sdk/platform-tools/` directory. One of the most common platform tools is adb:

> **adb (Android Debug bridge):** `adb` is a command-line tool that acts as a bridge between your machine and either an emulated device or a physical device at a given time. Using `adb`, you can install new applications to the mobile device, browse the log files of the current applications, and so on. You can find the `adb` tool in `sdk/platform-tools/`.

> **NOTE** *For more information about* `adb` *and how to use it, refer to* http://developer.android.com/guide/developing/tools/adb.html.

Other platform tools such as `aidl`, `aapt`, `dexdump`, and `dx` are called by Android built tools of the Android Development Tool (ADT). However, you rarely use them explicitly.

SDK TOOLS

SDK tools are installed by default when you install the Android SDK. If you are developing Android applications, you also need to use these tools, which are updated periodically. These tools are located in the `sdk/tools/` directory. The most important SDK tools are described here.

> **NOTE** *For more information about SDK tools and other available tools, refer to* `http://developer.android.com/guide/developing/tools/`.

DDMS: The Dalvik Debug Monitor Server (DDMS) is a debugging tool that provides various services for the Android, such as port-forwarding services, screen capture on the device, thread and heap information on the device, logcat, radio state information, incoming call and SMS spoofing, and location data spoofing. In Android OS, every application runs in its own process and, as usual, in its own VM. Hence, each VM exposes a unique port that a debugger can attach to.

DDMS can be used within Eclipse and also from the `sdk/tools/` directory. To run it from Eclipse, you should click Window ➪ Open Perspective ➪ DDMS. To run it from the command line, you need to go to the `sdk/tools/` directory and then type **ddms** (or **./ddms** on Mac/Linux).

android: `android` is a development tool that enables you to create, delete, or view AVDs. Using `android`, you can also update your Android SDK components. The `android` tool can be found in the `sdk/tools/` directory. You do not need to use this tool directly because its features are already integrated into Eclipse through the ADT plug-in. However, if you want to use it from the command line, you first need to go to the `sdk/tools/` directory and type **android** (or **./android** on Mac/Linux).

> **NOTE** *For more information on* `android`, *refer to* `http://developer.android` `.com/guide/developing/tools/android.html`.

emulator: `emulator` is a virtual mobile device that allows AVDs to run in an emulated window. It simply enables you to develop and test Android applications without using any actual mobile device. Using the `emulator`, you can interact with the emulated device as you would interact with the actual device.

> **NOTE** *For more information on Android* emulator, *refer to* http://developer
> .android.com/guide/developing/tools/emulator.html.

sqlite3: sqlite3 is a command-line shell that you can use to manage SQLite databases from Android applications.

> **NOTE** *For more information on* sqlite3, *refer to* http://developer.android
> .com/guide/developing/tools/adb.html#sqlite *and* http://www.sqlite
> .org/sqlite.html.

hierarchyviewer: Hierarchy Viewer enables you to debug and optimize the user interface. It provides a visual representation of the layout's View hierarchy (that is, Layout view) and a magnified inspector of the display (that is, Pixel Perfect view). The hierarchyviewer tool can be found in the sdk/tools/ directory. To run it from the command line, you first need to go to the sdk/tools/ directory and then type **hierarchyviewer** (or **./hierarchyviewer** on Mac/Linux).

> **NOTE** *For more information on the Hierarchy Viewer, refer to* http://devel-
> oper.android.com/guide/developing/tools/hierarchy-viewer.html.

zipalign: This archive alignment tool provides important optimization to Android application (.apk) files. The purpose is to ensure that all uncompressed data starts with a particular alignment relative to the start of the file. The zipalign tool can be found in the sdk/tools/ directory. To run it from the command line, you first need to browse the sdk/tools/ directory and then type **zipalign** (or **./zipalign** on Mac/Linux).

> **NOTE** *For more information on* zipalign, *refer to* http://developer
> .android.com/guide/developing/tools/zipalign.html.

draw9patch: This tool allows you to easily create a NinePatch graphic using a WYSIWYG editor. It can be found in the sdk/tools/ directory. To run it from the command line, you first need to go to the sdk/tools/ directory and then type **draw9patch** (or **./draw9patch** on Mac/Linux).

> **NOTE** *For more information on* draw9patch *and how to use it, refer to* http://
> developer.android.com/guide/developing/tools/draw9patch.html.

ANDROID VIRTUAL DEVICE

To test your applications in an emulator, you need to create an Android mobile phone emulator which is called an AVD. When the Android Development Tool is installed, no AVD is created automatically. To create an AVD, follow these steps:

1. In Eclipse, click Window ➪ Android Virtual Device Manager.

2. The Android Virtual Device Manager window should be open. Click New in the open window to open the Create New AVD window.

3. Input the properties of your AVD as follows (see Figure 3-10):

AVD Name: Give a new name to your virtual device.

Device: This field enables you to set some properties of the AVD. A drop-down list gives all devices known in the current SDK installation.

Target: You should select the API Level of the virtual device. It simply represents the installed Android OS version on the mobile phone. For example, if you are developing for a mobile phone that has an installed Android API Level 12, you may select API Level 12, or you can simply choose the highest API level.

SD Card: Additionally, you may create an SD card in your AVD.

4. To finish, click the Create AVD button at the bottom.

After you perform these steps, your first AVD is created. When running your application, you may test it in different appropriate API levels and in different screen-sized AVDs so that you can create additional AVDs in the future.

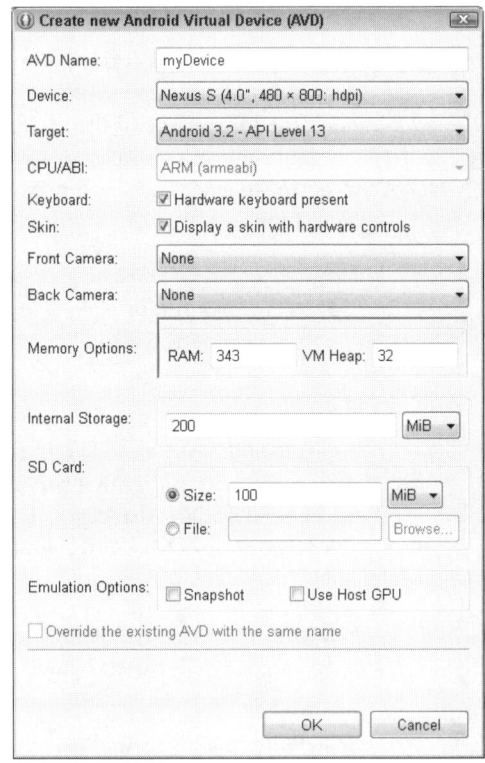

FIGURE 3-10

The screen size of some AVDs may be very large, so it might not fit into the screen of the development computer. To solve this issue, you should open AVD Manager ➪ AVD ➪ Start. In the open window, click Scale display to real size and then select the screen size in inches. Based on the selected screen size, the AVD's screen size is reduced or increased.

SUMMARY

Android is one of the most popular operating systems used in mobile phones. It is a modified version of Linux, owned by Open Handset Alliance and completely open sourced.

The operating system versions of Android are identified by API level numbers. When developing applications, you should consider the OS version of the target device, because some properties of the latest API levels might not have been introduced in the target version.

Android provides an SDK to write applications for Android mobile devices. The SDK provides a rich set of tools including debugger, libraries, mobile phone emulator with different API levels, documentation, sample codes, and tutorials. Android also provides an easy installation and IDE integration with a software package called ADT Bundle.

There are four different building blocks of Android applications; *activities*, *services*, *broadcast receivers* and *content* providers. Each android application uses at least one activity and may further use any number of other components as well. Furthermore, *intents and intent filters*, *manifest file*, and *applications resources* are used along with these components.

Manifest file is one of the most important components of Android applications. The device requirements, activities, and permissions should be defined in the manifest file for a successful installation. For example, in order to enable NFC service, the requirement of the NFC antenna in the mobile device must be defined in the manifest file. So, devices that do not have NFC capability cannot install your application.

Android uses a special virtual machine named DVM. This virtual machine uses a special byte code system called Dalvik executables developed specifically for Android executable files. When you develop an Android application, java compiler converts the source code into bytecode, and then dx converter converts it into Dalvik executables. Finally, the executable file is packed into an Android package file by using an archiver prior to installing it on a mobile phone. When the Eclipse IDE and ADT plug-in are used, the process to convert the code to an Android package file is automated.

Android Software Development Primer

➤ Creating your first Android application

➤ Running Android applications on mobile phones

➤ Distributing applications on Google Play

➤ Creating layouts for Android applications

➤ Using multiple layouts

➤ Using the event listener, linear layout, relative layout, grid layout, and dialog builders

➤ Implementing multiple activities

➤ Using menu items

WROX.COM CODE DOWNLOADS FOR THIS CHAPTER

The wrox.com code downloads for this chapter are found at www.wrox.com/remtitle .cgi?isbn=1118380096 on the Download Code tab. The code is in the Chapter 4 download and individually named according to the names throughout the chapter.

This chapter introduces application development in Android platforms for Java programmers. The chapter focuses specifically on how to develop, test, and distribute Android applications, mainly by incorporating commonly used Android APIs. The development environment is mainly the preferred IDE, which is Eclipse in this case; and the testing environments are emulators and mobile devices. This content is certainly sufficient to help Java programmers develop Android applications. Moderately skilled programmers can implement and run Android applications on both emulators and actual mobile phones after thoroughly studying and internalizing all the content here. The information in this chapter also provides sufficient background to understand the remaining chapters of the book and also to develop NFC-enabled Android applications afterward. Users who want to become more experienced Android programmers and pursue a proper career should use additional resources for that purpose after reading this book.

To understand this content, you must be able to create moderate-sized Java programs; however, no Android knowledge is required. If you do not have any experience on Android, you should start from Chapter 3, "Getting Started with Android"; there, you learn how to install the Android SDK and learn basic information with respect to being able to create and run Android projects before starting this chapter. If you have sufficient Android programming skills, you can skip this chapter and continue with Chapter 5, "NFC Programming: Reader/Writer Mode."

The Android developers' portal, `http://developer.android.com/`, is extremely useful; it is currently the resource most commonly used by Android developers around the world. We recommend that you navigate through this portal to get additional resources and up-to-date information.

The Android application development structure is based on the Java programming language. Therefore, we assume that you have the required Java knowledge. If you need to review, check out `http://docs.oracle.com/javase/tutorial/java/`, Oracle's site tutorial, for starters.

CREATING YOUR FIRST ANDROID APPLICATION

Before you can create an Android application, you first need to set up the environment as described in Chapter 3. The next step is to create an Android project. To do this, in Eclipse, select File ➪ New ➪ Project ➪ Android Application Project (see Figure 4-1). This opens a New Android Project Wizard. If you are using an existing sample or an existing source, select the related item.

To continue creating your new project, enter the following required information (see Figure 4-2):

➤ **Application Name:** This is the name that you want to assign to the application. When you install the application to the actual mobile device after development, this name appears in menu icons, application shortcuts, the title bar, and so on. For this example, we used `Hello World` for the application's name.

➤ **Project Name:** This is the name that you want to assign to your project.

➤ **Package Name:** Each Android package should have a unique name throughout the world. Therefore, you should be sure to follow this naming rule. The

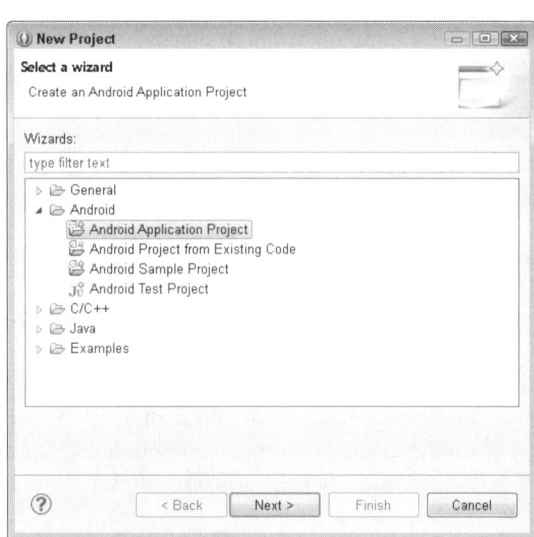

FIGURE 4-1

name generally consists of the publisher's domain name plus the application-specific name. In the `Hello World` example, we named the package `com.nfclab.helloworld`.

➤ **Minimum Required SDK:** This is the minimum version of the SDK that the application expects to execute. It generally should be the API level that you checked in the Build Target screen. We checked API level 3 in the Build Target screen, so the Minimum SDK for this application should be 3.

➤ **Target SDK:** In the Target SDK section, choose the Android platform and API level for which you are developing the project. Also, remember that because of the upward compatibility of Android applications, devices that use higher-level APIs can use your application, but it is not available for devices using lower-level APIs. Therefore, it is always safer to choose the earliest version that your application can run. For the Hello World application here, choose API level 3 because this API level is enough to run the application. If you have not installed API level 3 yet, you need to either install it or select a higher API level.

➤ **Compile With:** This is the platform version that the ADT will compile your application with. You will be able to use the features that this version provides.

➤ **Theme:** This part is the user interface style to apply for your application.

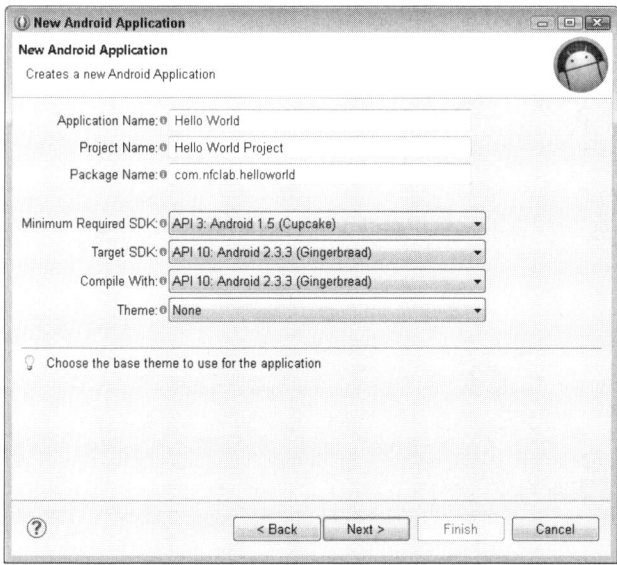

FIGURE 4-2

After you input the required data and click Next, you need to input the following data to configure your project (see Figure 4-3):

➤ **Create custom launcher icon:** If you wish to create a custom launcher icon for your application, you need to check the corresponding checkbox.

➤ **Create activity:** If you wish to create a new activity in your application, you need to check the corresponding checkbox.

➤ **Create Project in Workspace:**
Workspace is the folder in which your
project information and all related data
are saved. You may either choose the
default workspace location or any other
specific folder into which you want to
save project data.

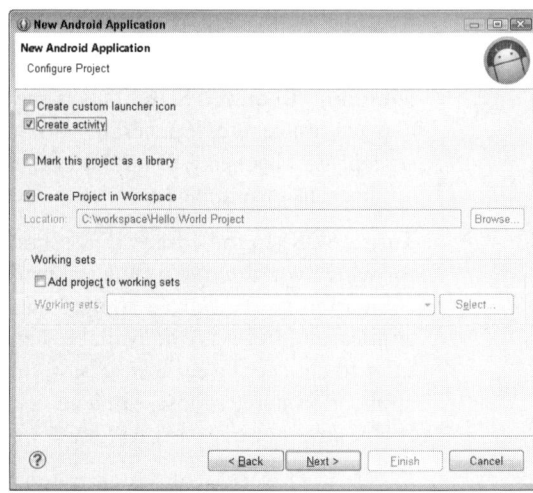

FIGURE 4-3

When you click Next, if you checked the
"Create custom launcher icon" checkbox,
you need to select an image file for your
application's launcher icon. If you checked
the "Create activity" checkbox, you will see
a Create Activity dialog box as shown in
Figure 4-4. In this screen you need to select
an activity type. The selected activity type
will automatically set up your activity's screen
layout, add code to the activity to handle layout
objects, and so on. For the first project, select BlankActivity as shown in Figure 4-4.

When you select the BlankActivity type and click Next, the name of the activity and layout are asked.
As described in Chapter 3, an activity is the user interface that the user interacts with. It is actually the
most basic component that the programmer uses for each project. Each project should have at least
one activity that interacts with the user as the application is invoked. The initial activity may further
call other activities, depending on the complexity and the design parameters of the application. An
activity is a Java class, so you have to name it using the same Java conventions you use for Java class
names. You may also specify a custom layout for the activity from the drop-down list as shown in
Figure 4-5. When you input the required data and click Finish, your project will be created.

FIGURE 4-4

FIGURE 4-5

Components of the Project

After you've created an Android project, Eclipse automatically adds the necessary components for that specific project and displays them in the Package Explorer on the left side of the window (see Figure 4-6).

The project components include the following:

➤ The source folder, `src`, consists of all the Java classes — activities, for example — created by the user.

➤ The `gen` folder holds the automatically generated Java files. Eclipse creates these files, so you should not modify them.

➤ `Android 2.3.3` is the Target SDK version selected for the project.

➤ The `assets` folder is the place where you can put some files and later retrieve them for the application.

(Package Explorer tree:)
- ▲ Hello World Project
  - ▷ src
  - ▷ gen [Generated Java Files]
  - ▷ Android 2.3.3
  - ▷ Android Dependencies
  - assets
  - ▷ bin
  - ▷ libs
  - ▲ res
    - ▷ drawable-hdpi
    - drawable-ldpi
    - ▷ drawable-mdpi
    - ▷ drawable-xhdpi
    - ▷ layout
    - ▷ menu
    - ▷ values
    - AndroidManifest.xml
    - proguard-project.txt
    - project.properties

FIGURE 4-6

➤ The `res` folder contains the resources, which include `drawable`, `layout`, and `values` folders by default:

  ➤ The `drawable` folders store the image files depending on the density of the screens.

  ➤ The `layout` folder contains the layouts that the project will use. When an Android project is created, a `main.xml` file is automatically generated for the layout of the activity. When you want to define additional screens for different reasons, you should define additional files in the same folder. Names of the layouts can consist of only lowercase letters. When you want to define new layout files, right-click Layout and then click New ➪ Other ➪ Android ➪ Android XML Layout File. Using different layouts in different activities is described later.

  ➤ The `values` folder holds the variable names and their value pairs, such as strings and colors. When the project is created, the `strings.xml` file is automatically created with two parameters. You can add all required values you want in the same file. All layout files can use the same file, `strings.xml`, and therefore no additional value file is necessary afterward.

  ➤ `AndroidManifest.xml`, as described in Chapter 3, consists of important information about the current project. If the application makes any attempt to access the Internet, the related code (intent filter) must be inserted into the manifest file first. There is a similar requirement for making calls. The XML file needs to declare all components of the project that are activities, services, content providers, and broadcast receivers. Also, it needs to contain the required permissions for the application. As you define activity class files, do not forget to register any activity into the manifest file.

Figure 4-7 shows the Android Manifest file. To open this file, simply double-click `AndroidManifest.xml` in Package Explorer. The screen shows the general manifest attributes, such as the package name, version code, and version name.

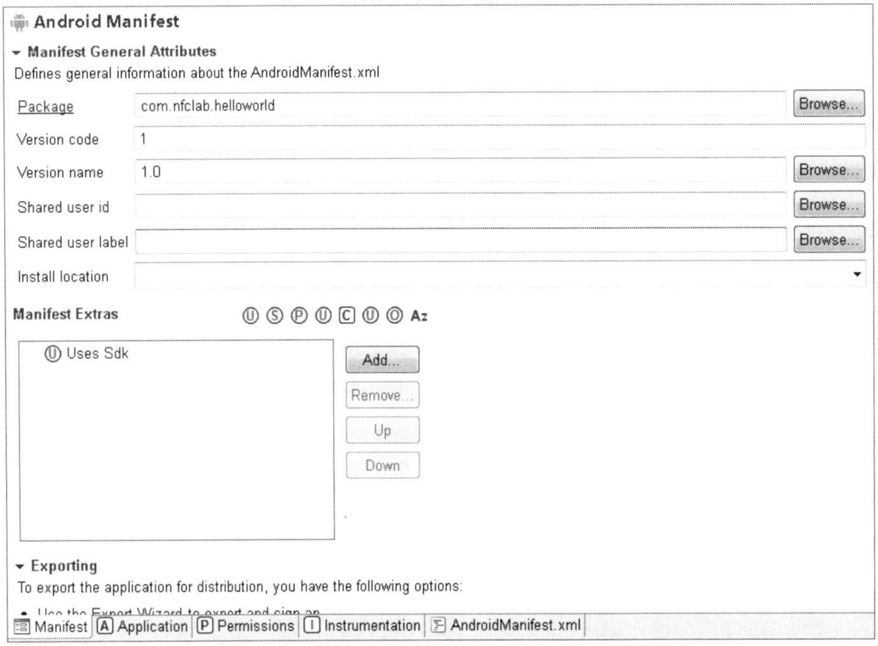

FIGURE 4-7

Version code contains important data to keep track of version numbers of applications to inform users about new versions. This capability is especially important for the applications uploaded to Google Play. For each version of an application, you should increment this number so that users are able to update the application to the next version from Google Play. When you first develop an application, its version code is initially set to 1. If you upgrade that application, the version number is incremented before it is made available so that the Android system understands that the application is upgraded by looking at the version number. When a new version of an existing

application is available, Android either downloads it automatically or informs users to download the new version, based on user preference. The version name is a string that gives information only about the version.

On the bottom of the window, you can see additional tabs for the manifest file, such as `Application`, `Permissions`, and `AndroidManifest.xml`. The `AndroidManifest.xml` tab at the bottom of the window displays the XML code for the manifest file. When you open the `AndroidManifest.xml` tab, you see the XML source for the file (see Figure 4-8). We generally use this tab to edit the manifest file.

FIGURE 4-8

In this file, the `package` attribute defines the base package name for the following Java elements. You can see that the version code and version name are also set in the `android:versionCode` and `android:versionName` attributes, respectively.

As you can see in Figure 4-8, the `uses-sdk` part of the `AndroidManifest.xml` file indicates the minimal version of the SDK that is valid for your application. The `<application>` tag provides the characteristics of your application:

➤ The `android:icon` element describes the icon of the application that is displayed in the Android applications menu. The `@drawable/ic_launcher` value refers to the `ic_launcher` `.png` file in the `res/drawable` folder.

➤ The `android:label` element describes the application name. The `@string/app_name` value refers to the `app_name` value in the resource files.

➤ The `<activity>` tag defines an `Activity` that points to the `HelloWorldActivity` class. An intent filter is registered for this class. This intent filter defines that this activity is started when the application starts with the `action android:name="android.intent.action` `.MAIN"`. Also, the category `category android:name="android.intent.category` `.LAUNCHER"` is defined for this intent filter; this category indicates that this application is added to the application directory on the Android device.

Running the Project

To run the project, click the Run button on the upper-left of the Eclipse window. When you run the project, Eclipse launches the emulator first and then the application, as described in previous sections. If you already have created an AVD as described previously, the application runs without asking any questions. However, if you have not created one yet, Eclipse requests that you create a new AVD. Refer to Chapter 3 to learn how to create a new AVD.

After running your application successfully, you should see output like that shown in Figure 4-9.

FIGURE 4-9

RUNNING APPLICATIONS ON YOUR MOBILE PHONE

There are two major options to test an application on your actual mobile phone. For the first option, while in the Eclipse environment, you may connect your mobile phone to a computer via a USB cable and immediately test it by running the application on the mobile device while it is still connected to the computer via the cable. The other option is manual installation by transferring the `.apk` file to the mobile and running the application inside the mobile.

Currently, the emulator does not simulate NFC, so you should run your NFC applications in your mobile device. For this reason, this step is important in NFC application development.

Running Applications Instantly

To run NFC-based Android applications instantly on your mobile phone, follow these steps:

1. Add the following code inside the `<application>` tag:

    ```
    <application android:debuggable="true" >
    ```

2. By default, your Android device does not allow the installation of applications from external sources. Allow installation by applying the following settings in the device. Go to Settings ⇨ Applications (for Android 4.0, go to Settings ⇨ Security) and enable Unknown Sources.

3. Turn on USB Debugging in your device. To do so, go to Settings ⇨ Applications ⇨ Development (for Android 4.0, go to Settings ⇨ Developer Options) and enable USB Debugging. Additionally, you need to enable the Stay Awake option to disable screen sleeping.

4. Set up your system to detect the mobile device as follows. If you are developing with Mac OS X, this step is not needed. If your development environment is Windows, install the USB driver of the mobile device to your computer to use the USB cable for data transfer. If your mobile phone is Nexus One or Nexus S, you can visit `http://developer.android .com/sdk/win-usb.html` to download the USB driver. Otherwise, you can visit `http:// developer.android.com/sdk/oem-usb.html`. Moreover, if your development environment is Linux, create a file named `/etc/udev/rules.d/51-android.rules` and then add the following line to this file:

    ```
    SUBSYSTEM=="usb", ATTR{idVendor}==" 04E8", MODE="0666", GROUP="plugdev"
    ```

> **NOTE** *The vendor ID given in this example is for Samsung. For other vendors' IDs, refer to* `http://developer.android.com/guide/developing/device .html#VendorIds`.

After completing these steps, you are ready to run your application on the actual mobile device. For this purpose, select Run ⇨ Run Configurations. In the Target tab, select Always Prompt to Pick Device under Deployment Target Selection Mode and then click Run. The Android Device Chooser menu opens, and you should see your connected Android mobile device in the menu. Select your device and then click the OK button. Then the application is installed to your device and run.

> **NOTE** *For more information on using hardware devices, refer to* `http://developer .android.com/guide/developing/device.html`.

Running Applications by Manual Installation

The second option to test an application on the mobile device is to transfer the `.apk` file to the mobile device, install it, and execute it on the mobile as explained previously. Follow these steps to run your application on the mobile device after transferring the `.apk` file:

1. By default, your Android device does not allow the installation of applications from external sources. Allow the installation by going to Settings ⇨ Applications (for Android 4.0, go to Settings ⇨ Security and enable Unknown Sources).

2. When you run your project in Eclipse, the required `.apk` file is created in the project's workspace's `bin` folder. Copy and paste the `.apk` file to your mobile phone storage.

3. Locate the file using a file manager, double-click the `.apk` file, and install it.

DISTRIBUTING ANDROID APPLICATIONS

To distribute your application to users, you need to first create a release-ready package that users can install and run on their Android-powered devices. The package includes the same components as the debug `.apk` file (for example, compiled source code, resources, manifest file). This package is built using the same build tools; however, it is signed with your own certificate and optimized with the zipalign tool. If you build your application on Eclipse with ADT plugging and an Ant build script, the signing and optimization processes are simple. You can use the Eclipse Export Wizard to compile, sign, and optimize the application.

This section summarizes how to distribute your Android applications to users and provides useful guidelines.

1. Gathering Materials and Resources

To release your application, you need to obtain the following supporting materials and resources for the application:

➤ **Cryptographic keys:** The application should be digitally signed with a certificate by the application's developer. It is important to establish trust between applications and identify the author of the application.

➤ **Application icon:** The application icon should meet the recommended icon guidelines. This icon helps users identify the application on the device's Home screen and in the Launcher window, as well as on Google Play.

➤ **End-user license agreement:** This agreement helps to protect personal, organizational, and intellectual property.

➤ **Miscellaneous materials:** You also need to prepare promotional and marketing materials, text, or screenshots to distribute the application effectively.

2. Configuring the Application for Release

After collecting all required materials, you can start to configure the application for release. You may need to perform some configurations on the source code, resource files, and application manifest file. Some of the following configurations are optional, and some of them may have already been performed during the development process:

➤ **Choose a nice package name:** The package name cannot be changed after distribution, even for new versions. Therefore, you should select the package name carefully.

➤ **Turn off logging and debugging:** You need to deactivate logging by removing calls to log methods in the source files, and you also need to disable debugging by removing `android:debuggable attribute` from the `<application>` tag or by setting the `android:debuggable attribute` to `false` in the manifest file before building the application for release.

➤ **Clean up project directories:** The project should conform to the directory structure described for Android projects. Doing this is essential because leaving stray or orphaned files in the project may lead to the application creating problems during execution or behaving unpredictably.

➤ **Review and update manifest settings:** You should ensure that some manifest items are set correctly — particularly the `<uses-permission>` element, `android:icon` and `android:label` attributes in the `<application>` element, and `android:versionCode` and `android:versionName` attributes. You should also set several additional manifest elements before releasing the application on Google Play, such as `android:minSdkVersion` and `android:targetSdkVersion` attributes in the `<uses-sdk>` element.

➤ **Address compatibility issues:** Android provides several tools to make the application compatible with various devices. You can:

➤ Add support for multiple screen configurations

➤ Optimize your application for Android 3.0 devices

➤ Consider using the Support Library

➤ **Update URLs for servers and services:** If the application accesses remote servers and services, you should use a production URL or path for the server or service.

➤ **Implement licensing for Google Play release:** You should also consider adding support for Google Play Licensing, especially for paid applications. This is actually an optional configuration. Licensing enables application developers to control access to the application based on whether the user has purchased it.

3. Compiling and Signing with Eclipse ADT

As mentioned previously, you can use the Eclipse Export Wizard to export a signed `.apk` file. The Export Wizard performs all the interaction with the Keytool and Jarsigner, and allows developers to sign the package using a GUI instead of performing the manual procedures to compile, sign, and align. After compiling and signing the package, the wizard performs package alignment with zipalign. At this point, your application is ready for distribution. To create a signed and aligned `.apk` in Eclipse, select the project in Package Explorer and then select File ➪ Export. Then open the Android folder and select Export Android Application ➪ Next. The Export Android Application wizard starts.

4. Publishing on Google Play

Google Play is a robust publishing platform that helps you publish, sell, and distribute Android applications throughout the world. Before publishing an application, you need to register as a Google Play developer. You need to create a developer profile, pay a registration fee, and agree to the Google Play Developer Distribution Agreement. After registration, you can access the Google Play Developer Console. This console enables you to upload applications, configure publishing options, and more. If you want to sell your applications, you also need to set up a Google Checkout Merchant account.

> **NOTE** *For more information on Android application publishing, refer to* `http://developer.android.com/guide/publishing/publishing.html`.

UNDERSTANDING HELLO WORLD

The first project, Hello World, consists of only one Java class, `HelloWorldActivity`. Listing 4-1 shows the code for this class, which is also the main activity of the project.

LISTING 4-1: Hello World Activity (Hello World Project\src\com\nfclab\helloworld\ HelloWorldActivity.java)

```
package com.nfclab.helloworld;
import android.os.Bundle;
import android.app.Activity;
import android.view.Menu;

public class HelloWorldActivity extends Activity {

    @Override
    protected void onCreate(Bundle savedInstanceState) {
        super.onCreate(savedInstanceState);
        setContentView(R.layout.main);
    }

    @Override
    public boolean onCreateOptionsMenu(Menu menu) {
        getMenuInflater().inflate(R.menu.main, menu);
        return true;
    }
}
```

The `HelloWorldActivity` class in this example extends the `Activity`. The `onCreate` method of the activity is invoked as the first statement. The `setContentView` method enables you to set the activity content from a layout resource, which has already been defined. `R.layout` `.main` refers to the `main.xml` resource file, which resides in the `res/layout` folder. Moreover, the `onCreateOptionsMenu` method is also defined automatically in the code, which gets the menu items from the `res/menu/main.xml` file and displays in the settings menu of the application. The layout and menu files are created automatically as the project is generated. The resources for an Android project, such as layout files, images, and so on, should be stored in the `res` folder of that project. Listing 4-2 gives the `main.xml` layout file.

LISTING 4-2: Hello World Layout Resource (Hello World Project\res\layout\main.xml)

```
<RelativeLayout xmlns:android="http://schemas.android.com/apk/res/android"
    xmlns:tools="http://schemas.android.com/tools"
    android:layout_width="match_parent"
    android:layout_height="match_parent"
    tools:context=".HelloWorldActivity" >

    <TextView
        android:layout_width="wrap_content"
        android:layout_height="wrap_content"
        android:layout_centerHorizontal="true"
        android:layout_centerVertical="true"
        android:text="@string/hello_world" />

</RelativeLayout>
```

Figure 4-10 illustrates how layouts are organized. `ViewGroup` includes different views and `ViewGroups`. The `ViewGroup` class provides a base for subclasses called layouts. These layouts also offer different kinds of layout architecture such as linear, tabular, and relative. Moreover, the `View` object stores layout parameters and content for a specific area of the project.

In the example, `ViewGroup` is defined using `RelativeLayout`, which orients the positions of each element in relation to each other. Other commonly used `ViewGroups` are `LinearLayout`, `AbsoluteLayout`, `AdapterView`, and `GridLayout`.

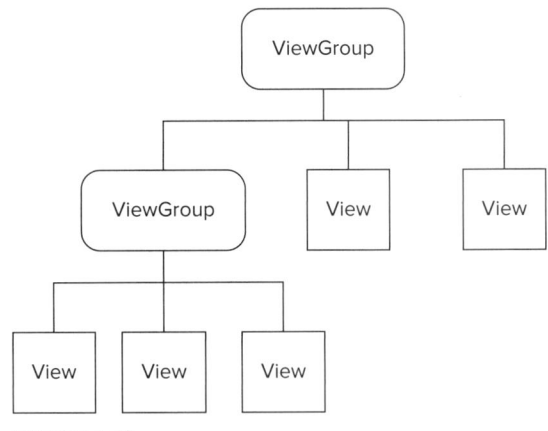

FIGURE 4-10

In the example, the `android:layout_width` and `android:layout_height` attributes specify the width and height of the view. The `match_parent` constant value means that the view can be as big as its parent, minus the parent's padding, if any. Another available constant value is `wrap_content`, which means that the view can be just large enough to fit its own internal content, taking its own padding into account.

Inside the `ViewGroup`, a `TextView` is created whose purpose is to display some text to the user. You can see that there are new elements in the tag. `android:layout_centerHorizontal` centers the view horizontally with its parent. On the other hand, `android:layout_centerVertical` centers the view vertically. The last element `android:text` describes the default text of the `TextView`. It is set to `@string/hello_world`, which indicates that the value is the `hello_world` constant's value, which is defined in `res/values/strings.xml` file. Instead of defining a constant, you can define the string for `TextView` in the XML file:

```
android:text="Hello world!"
```

Constants are stored in an XML file in the `res/values` folder of the project, but the name of the XML file does not matter. By default, constants are stored in the `strings.xml` file in the first example, which is shown in Listing 4-3.

LISTING 4-3: Hello World Strings Resource (Hello World Project\res\values\strings.xml)

```
<?xml version="1.0" encoding="utf-8"?>
<resources>

    <string name="app_name">Hello World</string>
    <string name="hello_world">Hello world!</string>
    <string name="menu_settings">Settings</string>
</resources>
```

As you can see from Listing 4-3, ADT automatically defines three strings in the file when creating the project. The `hello_world` variable is further used in the `main.xml` layout file, whereas the `app_name` variable is used in the manifest file and `menu_settings` is used in the `res/values/main.xml` to give a name to the menu item.

To access the user interface elements in the code such as `hello` and `app_name`, add identifier attributes in the layout file. The following line adds the ID of `myTextView` to the `TextView`:

```
android:id="@+id/myTextView"
```

You can then access it in the code by using the `findViewById` method:

```
TextView myTextView=(TextView) findViewById(R.id.myTextView);
```

You have two options to create layouts. The first one is to define them in XML files, and the other option is to define them in the program code.

An example to create layouts inside the code is shown here. As you can see, the `setContentView` method is not called to set the layout to an XML file in this case; `LinearLayout` and `TextView` are created instead. A `TextView` object is then added to the `LinearLayout` and is set as the content:

```
public void onCreate(Bundle savedInstanceState) {
    super.onCreate(savedInstanceState);
    LinearLayout linearLayout = new LinearLayout(this);
    linearLayout.setOrientation(LinearLayout.VERTICAL);
    TextView myView = new TextView(this);
    myView.setText("Hello World!");
    linearLayout.addView(myView);
    setContentView(linearLayout);
}
```

Using XML files is a better approach and the preferred option to create layouts instead of using code. Throughout the book, we create layouts in XML files.

USING MULTIPLE VIEWS

You will also use multiple views in the layout file. In the following code snippet, three different `TextViews` are created with different IDs and different default texts. Additional properties of the `TextViews`, such as text size and text color are also defined in the code. You can discover many other properties for a layout object online at `http://developer.android.com/reference` or by adding `"android:"` inside the `<TextView>` tag in the XML file. Eclipse displays the available properties for the layout object:

```xml
<?xml version="1.0" encoding="utf-8"?>
<LinearLayout xmlns:android="http://schemas.android.com/apk/res/android"
 android:layout_width="fill_parent"
 android:layout_height="fill_parent"
 android:orientation="vertical" >
    <TextView
     android:id="@+id/hello1"
     android:layout_width="fill_parent"
     android:layout_height="wrap_content"
     android:text="@string/helloWorld1"
     android:textSize="12pt"/>
    <TextView
     android:id="@+id/hello2"
     android:layout_width="fill_parent"
     android:layout_height="wrap_content"
     android:text="@string/helloWorld2"
     android:textSize="10pt" />
    <TextView
     android:id="@+id/hello3"
     android:layout_width="fill_parent"
     android:layout_height="wrap_content"
     android:text="@string/helloWorld3"
     android:textColor="@color/red"/>
</LinearLayout>
```

The layout file defines four different constants: `@string/helloWorld1`, `@string/helloWorld2`, `@string/helloWorld3`, and `@color/red`. These constants should be defined in an XML file in the `res/values` folder. The required code snippet is shown here:

```xml
<?xml version="1.0" encoding="utf-8"?>
<resources>
    <string name="app_name">Hello World</string>
    <string name="helloWorld1">Hello World </string>
    <string name="helloWorld2">Hello World again with smaller font</string>
    <string name="helloWorld3">Hello World again in default font,
                       but in red color </string>
    <color name="red">#FF0000</color>
</resources>
```

ANDROID PROJECT RESOURCES

In Android projects, you should externalize your application resources such as images, colors, and strings from your source code so that you can manage them independently. Notice that the Model-View-Controller (MVC) concept also is satisfied here. Moreover, it is important to provide alternative resources for specific device configurations.

After externalizing application resources, you can access them using resource IDs that are generated in the project's R class. Table 4-1 summarizes the resource directories supported inside the project and provides alternative resources for specific device configurations.

TABLE 4-1: Resource Directories Supported Inside the Project `res/` Directory

DIRECTORY	RESOURCE TYPE
`animator/`	XML files that define property animations.
`anim/`	XML files that define tween animations.
`color/`	XML files that define a state list of colors.
`drawable/`	XML files or Bitmap files such as `.png`, `.jpg`, `.gif` that are compiled into the drawable subtypes (that is, bitmap files, nine patches, shapes, animation drawables, state lists, and other drawables).
`layout/`	XML files that define a user interface layout.
`menu/`	XML files that define application menus that are described in the "Using Menu Items" section later in this chapter.
`raw/`	A directory for saving arbitrary files.
`values/`	XML files that contain simple values such as strings, integers, and colors. Files in the `values/` directory describe multiple resources. For a file in this directory, each child of the `<resources>` element defines a single resource. For example, a `<string>` element creates an `R.string` resource, and a `<color>` element creates an `R.color` resource.
`xml/`	Arbitrary XML files and configuration files that can be read at run time through `Resources.getXML()`.

Alternative Resources

Every application needs to provide alternative resources to support specific device configurations. For example, the application may include alternative drawable resources for different screen resolutions as well as alternative string resources for different languages. At run time, the Android loads the appropriate resources for the application based on these alternative resources.

To specify alternative resources for specific device configurations, you first need to create a new directory named `<resources_name>-<config_qualifier>` in the res folder. Here, `<resources_name>` is the directory name of the corresponding default resources, and `<qualifier>` is a name that specifies an individual configuration for which these resources are to be used. You can add more than one `<qualifier>`. Then you should save the respective alternative resources in the newly created directory. You need to name the resource files exactly the same as the default resource files.

> **NOTE** *For more information on providing project resources, refer to* `http://developer.android.com/guide/topics/resources/providing-resources.html`.

Accessing Resources

After creating the related resource files, you can use them either in XML files or in the code. You may refer to the resources in the code to place them in the layout or make modifications to them. To use the resources in the code, refer to them as follows:

```
R.<resource_type>.<resource_name>
```

Following are some examples for accessing resources in your code:

➤ `R.string.table_title`, which calls a string value from the resource file

➤ `R.drawable.background_image`, which calls a drawable image

To use the resources in XML files, refer to them through `@<resource_type>/<resource_name>`.

Here are some examples for accessing resources in XML files:

➤ `@color/blue`, which calls a color value from the resource file

➤ `@string/hello`, which calls a string value from the resource file

USING AN EVENT LISTENER

In the following example, you build a miles-to-kilometers converter, which converts the inputted value based on the selected converter type. The application is shown in Figure 4-11.

Layout

The layout file for this example has two radio buttons to select the converter, an `EditText` to input a value, a `Button` to submit the value, and a `TextView` to display the result. The layout XML file for the application is shown in Listing 4-4.

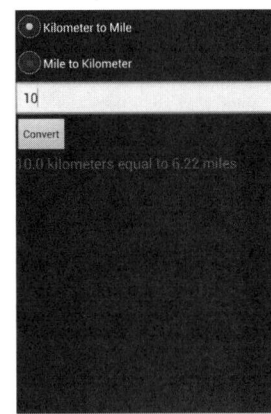

FIGURE 4-11

LISTING 4-4: Miles-to-Kilometers Converter Layout Resource (MileAndMeterConverter\res\layout\main.xml)

```xml
<?xml version="1.0" encoding="utf-8"?>
<LinearLayout xmlns:android="http://schemas.android.com/apk/res/android"
 android:orientation="vertical"
 android:layout_width="fill_parent"
 android:layout_height="fill_parent">

<RadioGroup
```

```
      android:id="@+id/radioGroup1"
      android:layout_width="wrap_content"
      android:layout_height="wrap_content"
      >
         <RadioButton
          android:id="@+id/radio0"
          android:layout_width="wrap_content"
          android:layout_height="wrap_content"
          android:checked="true"
          android:text="@string/meter"/>
         <RadioButton
          android:id="@+id/radio1"
          android:layout_width="wrap_content"
          android:layout_height="wrap_content"
          android:text="@string/mile"
         />
    </RadioGroup>

    <EditText
     android:id="@+id/inputField"
     android:layout_width="fill_parent"
     android:layout_height="wrap_content"
     android:inputType="numberDecimal|numberSigned" >
     <requestFocus/>
    </EditText>

    <Button
     android:id="@+id/convertButton"
     android:layout_width="wrap_content"
     android:layout_height="wrap_content"
     android:onClick="onClickHandler"
     android:text="@string/convert"
    />

    <TextView
     android:id="@+id/outputText"
     android:layout_width="fill_parent"
     android:layout_height="wrap_content"
     android:textColor="#cc0000"
     android:textSize="9pt"
    />

</LinearLayout>
```

> **NOTE** *The IDs in the XML file are further used in codes to retrieve the contents of the elements and change their contents if needed.*

As you can see from the layout file, the screen is designed using LinearLayout. Two RadioButtons are defined inside a RadioGroup. Defining the radio buttons in a group ensures that when one radio button is checked, all other radio buttons are automatically unchecked.

Following the buttons, the `EditText` element is created; this element enables users to input values from a text box. In addition, the `<requestFocus/>` element gives initial focus to the `EditText` element on the screen. If you want users to input only numeric values, add the following line inside the `EditText` element:

```
android:inputType="numberDecimal|numberSigned"
```

The Convert button implements an event handler using `android:onClick`. So, when the button is clicked, the `onClickHandler` method is invoked; it is defined as a value in the `onClick` attribute in the layout file. Finally, a `TextView` element is defined to display the results on the screen.

> **NOTE** *The* `android:onClick` *attribute is added in API level 4 (Android 1.6). Thus, you should create your project with at least API level 4.*

Resources

Resources are created in an XML file in the `res/values` folder. Listing 4-5 shows the implemented resources. Note that these resources are used in the layout file.

LISTING 4-5: Miles-to-Kilometers Converter Strings Resource (MileAndMeterConverter\res\ values\strings.xml)

```xml
<?xml version="1.0" encoding="utf-8"?>
<resources>
    <string name="app_name">Mile And Meter Converter</string>
    <string name="mile">Mile to Kilometer</string>
    <string name="meter">Kilometer to Mile</string>
    <string name="convert">Convert</string>
</resources>
```

Code

`RadioButton`, `EditText`, and `TextView` objects are created inside the main class to retrieve the contents of the views when a user submits the form:

```java
public class MileAndMeterConverter extends Activity {
    private RadioButton meterToMileButton;
    private RadioButton mileToMeterButton;
    private EditText input;
    private TextView output;
```

Inside the `onCreate` method, the content is set to the `main.xml` file. The objects, which are created in the main class, are initialized with the layout views. You retrieve the contents of the views later in the code by using methods such as `getText()` and `isChecked()`. For example, the input from the user is identified as `R.id.inputField`, and you get this value and save it as `input` for ease of use:

```
        @Override
        public void onCreate(Bundle savedInstanceState) {
        super.onCreate(savedInstanceState);
        setContentView(R.layout.main);
        meterToMileButton = (RadioButton) findViewById(R.id.radio0);
        mileToMeterButton = (RadioButton) findViewById(R.id.radio1);
        input = (EditText) findViewById(R.id.inputField);
        output = (TextView) findViewById(R.id.outputText);
    }
```

The onClickHandler method is invoked when the user presses the Convert button (see following code). The current view is then passed to the method as a parameter. The view.getId() method returns the current view's identifier. The ID of the current view is checked if the Convert button is pressed. Then the length of the input value for EditText is checked. If the length is 0, the application requests the user to enter a new number; otherwise, the checked RadioButton is compared to determine the convert operation that the user requested. If the Kilometer To Mile box is checked, the result is obtained from the meterToMile method. Otherwise, the result is obtained from the mileToMeter method. The result is displayed by setting the value of the TextView with the setText method:

```
public void onClickHandler(View view) {
    if(view.getId() == R.id.convertButton){
        if (input.getText().length() == 0) {
            output.setText("Please enter a number");
            return;
        }
        float inputValue = Float.parseFloat(input.getText().toString());
        if (meterToMileButton.isChecked()) {
            output.setText(inputValue + " kilometers equal to " +
                String.format("%.2f", meterToMile(inputValue)) + " miles");
        } else if(mileToMeterButton.isChecked()){
            output.setText(inputValue+" miles equal to " +
                String.format("%.2f", mileToMeter(inputValue)) + " kilometers");
        }
    }
}
```

The meterToMile and mileToMeter methods simply get the parameter (kilometer or mile) and return the result of the conversion, as shown here:

```
private float meterToMile(float kilometer) {
    return (float) (kilometer/1.609);
}

private float mileToMeter(float mile) {
    return (float) (mile*1.609);
}
```

USING RELATIVE LAYOUT

In this example, the previous example is modified to create a relative layout. In relative layouts, the positions of the views are defined in relation to each other and are placed on the screen accordingly. The screen view of this example is the same as the previous example (refer to Figure 4-11), and the layout file is shown in Listing 4-6. Hence, both layout options display the same visual effect, except that different layout codes are used in the background layout code.

LISTING 4-6: RelativeLayout Layout Resource (MileAndMeterConverter_Relative\res\layout\main.xml)

```xml
<?xml version="1.0" encoding="utf-8"?>
<RelativeLayout xmlns:android="http://schemas.android.com/apk/res/android"
 android:layout_width="fill_parent"
 android:layout_height="fill_parent">

<RadioGroup
 android:id="@+id/radioGroup1"
 android:layout_width="fill_parent"
 android:layout_height="wrap_content"
>
    <RadioButton
     android:id="@+id/radio0"
     android:layout_width="wrap_content"
     android:layout_height="wrap_content"
     android:checked="true"
     android:text="@string/meter"
    />
    <RadioButton
     android:id="@+id/radio1"
     android:layout_width="wrap_content"
     android:layout_height="wrap_content"
     android:text="@string/mile"
    />
</RadioGroup>

<EditText
 android:id="@+id/inputField"
 android:layout_width="fill_parent"
 android:layout_height="wrap_content"
 android:inputType="numberDecimal|numberSigned"
 android:layout_below="@+id/radioGroup1" >
 <requestFocus/>
</EditText>

<Button
 android:id="@+id/convertButton"
 android:layout_width="wrap_content"
 android:layout_height="wrap_content"
 android:text="@string/convert"
 android:onClick="onClickHandler"
 android:layout_below="@+id/inputField"
/>

<TextView
 android:id="@+id/outputText"
 android:textSize="9pt"
```

```
    android:textColor="#cc0000"
    android:layout_width="fill_parent"
    android:layout_height="wrap_content"
    android:layout_below="@+id/convertButton"
/>

</RelativeLayout>
```

As you can see in the listing, a relative layout is defined with the `<RelativeLayout>` tag. The most significant difference for this example is the `android:layout_below` attributes, which position the top edge of the view below the given view. For example, `EditText` is positioned below the radio group with an ID of `radioGroup1` by defining its layout as:

```
    android:layout_below="@+id/radioGroup1"
```

> **NOTE** *More attributes are available for relative layouts. For more information, refer to* `http://developer.android.com/reference/android/widget/RelativeLayout.html`*.*

Similar to `android:layout_below`, `android:layout_above` positions the bottom edge of the view above the given view. Both `android:layout_toRightOf` and `android:layout_toLeftOf` work in a similar way and likewise position the view's left or right edge to the right or left of the given view.

USING DIALOG BUILDERS

Dialogs are the small windows that appear in front of an activity and gain focus for user input. In this section, the example given in the "Using Relative Layout" section is modified to provide a better-organized user interface using an additional Exit option (see Figure 4-12) and an alert dialog box (see Figure 4-13).

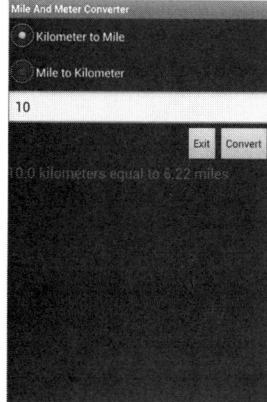

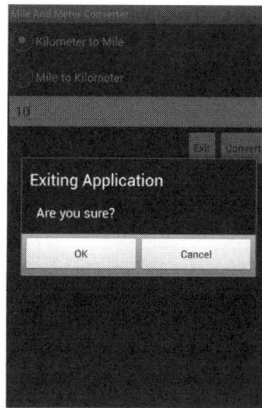

FIGURE 4-12   **FIGURE 4-13**

Layout

To provide a better user interface, you edit the layout file of the example given in the "Using Relative Layout" section. Everything remains the same except the Convert button. Additionally, you set the Cancel button to provide an option for the user to quit the application.

You modify the Convert button and set `layout_alignParentRight` to `true`, which makes the right edge of the button match with the right edge of the `EditText`:

```
<Button
 android:id="@+id/convertButton"
 android:layout_width="wrap_content"
 android:layout_height="wrap_content"
 android:text="@string/convert"
 android:onClick="onClickHandler"
 android:layout_below="@+id/inputField"
 android:layout_alignParentRight="true"
/>
```

An additional Cancel button is added afterward. Also, an `onClick` attribute is added that invokes the `onClickHandler` method when the Cancel button is pressed. You align this button to the left of the Convert button by setting the `layout_toLeftOf` attribute. You also align the top of this button the same as the Convert button by setting the `layout_alignTop` attribute. These two attributes enable you to place these buttons beside each other:

```
<Button
 android:id="@+id/cancelButton"
 android:layout_width="wrap_content"
 android:layout_height="wrap_content"
 android:text="@string/exit"
 android:onClick="onClickHandler"
 android:layout_toLeftOf="@+id/convertButton"
 android:layout_alignTop="@+id/convertButton"
/>
```

Resources

In the resource file, a string resource is added, which is used for the Exit button's `text` attribute in the layout file:

```
<string name="exit">Exit</string>
```

Code

In the source code, you need to add the required codes to handle pressing the Cancel button. When the Cancel button is pressed, the application opens an alert dialog asking whether the user actually wants to exit the application (see Figure 4-13). To perform this action, you modify the `onClickHandler` method, because the Cancel button's `onClick` attribute also invokes this method in the same way as it does for the Convert button:

```
public void onClickHandler(View view) {
    if(view.getId() == R.id.cancelButton){
        AlertDialog.Builder newAlert = new AlertDialog.Builder(this);
        newAlert.setTitle("Exiting Application");
```

```
        newAlert.setMessage("Are you sure?");
        newAlert.setPositiveButton("OK", new DialogInterface.OnClickListener(){
          public void onClick(DialogInterface dialog, int which) {
            MileandMeterConverterwithCancel.this.finish();
          }
        });

        newAlert.setNegativeButton("Cancel", new
                                   DialogInterface.OnClickListener() {
          public void onClick(DialogInterface dialog, int which) {
            dialog.cancel();
          }
        });

        newAlert.show();
    } else if(view.getId()==R.id.convertButton){
      if (input.getText().length() == 0) {
        output.setText("");
        Toast.makeText(this,
                   "Please enter a number",Toast.LENGTH_LONG).show();
        return;
      }
      float inputValue = Float.parseFloat(input.getText().toString());
      if (meterToMileButton.isChecked()) {
        output.setText(inputValue + " kilometers equal to " +
          String.format("%.2f", meterToMile(inputValue)) + " miles");
      } else if(mileToMeterButton.isChecked()) {
        output.setText(inputValue + " miles equal to " +
          String.format("%.2f", mileToMeter(inputValue)) + " kilometers");
      }
    }
  }
}
```

Because click events of two buttons, namely OK and Cancel, are handled in the same event handler, the same event handler is invoked when either of the two buttons is pressed. Hence, you should understand which button is clicked to perform required actions in the event handler method. When the event handler is invoked, the reference to the View component that is clicked is sent to the method as a parameter. You actually compare that reference with the two button components, namely OK and Cancel, to understand which button is pressed. More specifically, you compare the view.getId() method with R.id.cancelButton because the ID of the Cancel button is cancelButton. Then an AlertDialog named newAlert is built with title text, message text, and two buttons. One of the buttons is the PositiveButton, which finishes the activity, and the other one is the NegativeButton, which cancels the exit process. Also, listeners are added to these buttons to get the user input. Finally, the alert is shown using the newAlert.show() method.

After performing the Cancel button operation, you check whether the pressed button is the Convert button. The remainder of the code operates in the same way.

To provide a better user interface, you may add a `Toast` to the warning message. A `Toast` is a view that contains a small message for the user and shows it for some period of time. To display a `Toast`, modify the related code as follows:

```
if (input.getText().length() == 0) {
    output.setText("");
    Toast.makeText(this, "Please enter a number",
    Toast.LENGTH_LONG).show();
    return;
}
```

USING GRID LAYOUT

In this section you learn how to use the grid layout. `GridLayout` is a layout option that places elements in a rectangular grid and has been available since API level 14. A number of available rows and columns are defined initially, and the coordinate of the item's cell is explicitly given in the layout file to place that component in the desired cell. The width of each column is resized automatically to encapsulate all items in that column. Therefore, inserting a wider component than the previous components in that column automatically increases the width of that column. You can use `columnSpan` to enable one single component to occupy more than one cell in the same row. Remember that in `LinearLayout` there is no width-sizing problem because all items are placed one after the other in one row. In `GridLayout`, `columnSpan` may be required to adjust the items. When compared to `LinearLayout`, `GridLayout` requires more effort to handle the positions, but it also is more flexible. Indexes of rows and columns start with 0. You can leave any cell empty, so you do not need to fill all cells in the grid.

Here is the good news from Android that follows the MVC (Model-View-Controller) concept: you need only the View part, the layout XML file, to change the display. Because the M and C parts are the same, you do not even open the related files.

Consequently, you can modify the previous example's layout XML file to see how to use the grid layout. The application is shown in Figure 4-14. Because `GridLayout` was introduced starting in API level 14, you should also change the API level of the application and set it to an API level of 14 or higher because the application won't work otherwise.

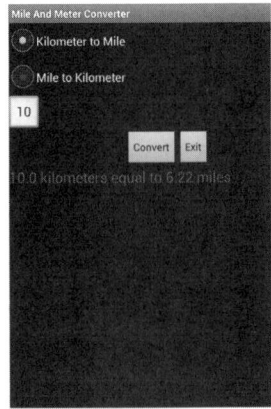

The layout file's source is given in Listing 4-7. Inside the `GridLayout` element are some attributes that were not used previously. The `android:alignmentMode` element is set to `alignBounds` so that the component is aligned between the edges of the view. The default value is `alignMargins`, which sets the alignment between the outer boundaries of a view. The `android:rowCount` and `android:columnCount` attributes define the number of available rows and columns in the grid. The indexes of the components should be {from 0 to rowcount-1} and {from 0 to columncount-1}, respectively. If the boundaries are exceeded, the application does not run. Also note that the index starts from 0, which means that four indices represent the indices 0, 1, 2, and 3.

FIGURE 4-14

LISTING 4-7: GridLayout Layout Resource (MileAndMeterConverter_Grid\res\layout\main.xml)

```xml
<?xml version="1.0" encoding="utf-8"?>
<GridLayout    xmlns:android="http://schemas.android.com/apk/res/android"
 android:layout_width="wrap_content"
 android:layout_height="wrap_content"
 android:alignmentMode="alignBounds"
 android:rowCount="4"
 android:columnCount="3"
>

<RadioGroup
 android:id="@+id/radioGroup1"
 android:layout_width="wrap_content"
 android:layout_height="wrap_content"
 android:layout_row="0"
 android:layout_column="0"
>
    <RadioButton
    android:id="@+id/radio0"
    android:layout_width="wrap_content"
    android:layout_height="wrap_content"
    android:checked="true"
    android:text="@string/meter"
    />
    <RadioButton
     android:id="@+id/radio1"
     android:layout_width="wrap_content"
     android:layout_height="wrap_content"
     android:text="@string/mile"
     />
</RadioGroup>

<EditText
 android:id="@+id/inputField"
 android:layout_height="wrap_content"
 android:layout_row="1"
 android:layout_column="0"
 android:inputType="numberDecimal|numberSigned" >
 <requestFocus/>
</EditText>

<Button
 android:id="@+id/convertButton"
 android:layout_width="wrap_content"
 android:layout_height="wrap_content"
 android:layout_row="2"
 android:layout_column="1"
 android:text="@string/convert"
 android:onClick="onClickHandler"
/>

<Button
 android:id="@+id/cancelButton"
```

continues

LISTING 4-7 *(continued)*

```
    android:layout_width="wrap_content"
    android:layout_height="wrap_content"
    android:layout_row="2"
    android:layout_column="2"
    android:text="@string/exit"
    android:onClick="onClickHandler"
/>

<TextView
    android:id="@+id/outputText"
    android:layout_row="3"
    android:layout_column="0"
    android:layout_columnSpan="3"
    android:textSize="9pt"
    android:textColor="#cc0000"
/>

</GridLayout>
```

In this example, four row and three column indices are created. After defining each index, you need to add `android:layout_row` and `android:layout_column` attributes to each element. These attributes set each element's position between indices.

For example, a `Button` element with the ID `convertButton` is in the second index of the row and in the first index of the column. An `EditText` element with the ID `inputField` is in the first index of the row and in the zeroth index of the column.

ANDROID ACTIVITY LIFECYCLE

In an Android system, activities are managed in a stack. When an activity starts, it moves to the front of the activity stack and becomes visible. The other running activities still run if you do not quit them, but they stay in lower levels in the stack and thus become invisible. When you stop the current activity, the previous activity returns to the front.

An activity mainly has four important states, which are illustrated in Figure 4-15:

➤ **Active:** An activity is active when it is at the top of the activity stack.

➤ **Paused:** An activity is paused when it is visible but another activity is at the top of the stack with a non-full-sized or transparent window. An activity in this state still maintains the information but can be killed when the Android OS requires memory for some other process.

➤ **Stopped:** An activity is stopped when another activity is started and moves to the front of the stack. An activity in this state is not visible and often may be killed when memory is needed elsewhere.

➤ **Destroyed:** An activity is destroyed when the system asks it to finish its process or kills its process. When the activity is launched, it is started from scratch.

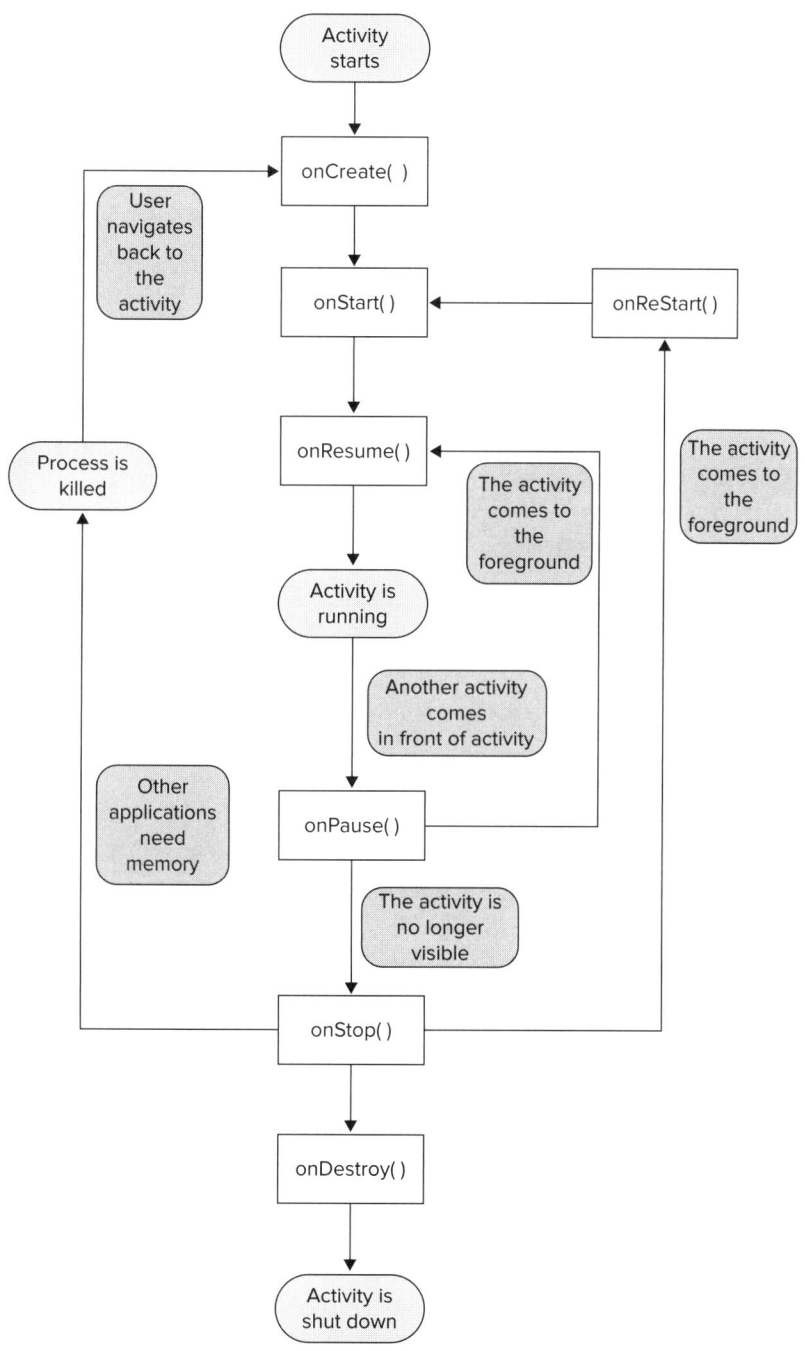

FIGURE 4-15

Table 4-2 describes the methods used along with the activity lifecycle. The table includes information on when an activity is called and which method may follow the current method.

> **NOTE** *For more information on the activity lifecycle, refer to* `http://developer.android.com/reference/android/app/Activity.html`.

TABLE 4-2: Methods in the Activity Lifecycle

METHOD	WHEN IT IS CALLED	FOLLOWED BY
onCreate()	When an activity is initially created	Always onStart()
onStart()	When an activity is becoming visible, which includes the time after it is created and the time it is coming back to the screen after being stopped	onResume() when the activity moves to the front of the activity stack onStop() when the activity becomes hidden
onRestart()	When an activity is stopped and before it is started again	Always onStart()
onResume()	When an activity starts interacting with the user	Always onPause()
onPause()	When another activity starts resuming	onResume() when the activity returns to the front onStop() when the activity becomes invisible to the user
onStop()	When the activity is no longer visible	onRestart() when the activity is coming back to interact with the user onDestroy() when the activity exits
onDestroy()	When the activity is destroyed	Nothing

IMPLEMENTING MULTIPLE ACTIVITIES AND INTENTS

In this section, you create an application that consists of multiple activities. The main activity displays a list, which consists of three options: Mile & Meter Converter, Celsius & Fahrenheit Converter, and Foot & Yard Converter, as shown in Figure 4-16. When the application is executed, the list is displayed. When a user selects an option from the list, the related activity comes to the front, as shown in Figure 4-17. A total of four activities are defined in this example: three activities to handle each selection and one activity to display the menu items and call the requested activity.

FIGURE 4-16

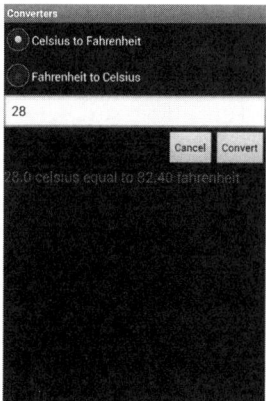

FIGURE 4-17

Step 1: Creating the Layout File

In the main activity, you display a list of the options (refer to Figure 4-16). In the layout, the description for the list items should be given as shown in Listing 4-8. Each list item's text size is 16sp and padding is 10dp.

> **NOTE** *Be aware that* sp *stands for scale-independent pixel, and* dp *stands for density-independent pixel. You can find more information on dimension values at* http://developer.android.com/guide/practices/screens_support .html

LISTING 4-8: ListView Layout Resource (MultipleActivities\res\layout\main.xml)

```
<?xml version="1.0" encoding="utf-8"?>
<TextView xmlns:android="http://schemas.android.com/apk/res/android"
 android:layout_width="fill_parent"
 android:layout_height="fill_parent"
 android:padding="10dp"
 android:textSize="16sp" >
</TextView>
```

Step 2: Building the ListView

In the main class file, you display a list to the user and also implement onClickListener to listen to user input and react accordingly. In the onCreate method, you implement a setListAdapter method, which automatically adds a ListView, as shown here:

```
@Override
public void onCreate(Bundle savedInstanceState) {
    super.onCreate(savedInstanceState);
    setListAdapter(new ArrayAdapter<String>(this, R.layout.main, CONVERTERS));
    ListView myList = getListView();
```

```
        myList.setTextFilterEnabled(true);
        myList.setOnItemClickListener(this);
    }

    static final String[] CONVERTERS = new String[] {
        "Mile & Meter Converter",
        "Celsius & Fahrenheit Converter",
        "Foot & Yard Converter"
    };
```

The method takes the `ArrayAdapter`, which is a special kind of `ListAdapter` that supplies data to `ListView`. The `ListView` provides a nice user interface for displaying the list of elements one by one, and `ListAdapter` supplies the required data to `ListView`. Thus, `ArrayAdapter` manages the list items that will be placed in `ListView`. The `ArrayAdapter` constructor takes the current application context, the resource ID for the layout file (`R.layout.main`) containing a `TextView` to use for each list item, and objects to be represented in the `ListView` (`CONVERTERS` array). Despite the fact that you have a layout file, you do not need the `setContentView` method because the layout file loads the properties of each list item.

The `getListView` method saves the built `ListView` to a `ListView` object named `myList`, and then `OnItemClickListener` is added via the `myList.setOnItemClickListener` method to get the user's selection on the list.

Step 3: Implementing onItemClick

When a user selects one of the items in the list, the `onItemClick` method is invoked automatically. The position of the selected item in the list is sent to the `onItemClick` method via a parameter named `position`. Using the position integer, you can determine the selected item from the list.

To start a new activity based on the user's selection, you need to use `Intents`. Remember that `Intents` are normally used to trigger an already-existing service available in the current device. When the user selects `Mile & Meter Converter` with a position of 0, a new intent is created from the `MileToMeter` class and is started using the `startActivity` method as shown here:

```
public void onItemClick(AdapterView<?> parent, View view, int position, long id) {
    if(position==0){
        Intent intent = new Intent(this, MileToMeter.class);
        startActivity(intent);
    } else if(position==1){
        Intent intent = new Intent(this, CelsiusToFahrenheit.class);
        startActivity(intent);
    } else if(position==2){
        Intent intent = new Intent(this, FootToYard.class);
        startActivity(intent);
    }
}
```

Step 4: Editing AndroidManifest.xml

All activities defined in the current project must be properly defined in the `AndroidManifest.xml` file of the project, as described earlier. The minimum definition of each additional activity should be as follows:

```
<activity android:name="CLASS_NAME"></activity>
```

Since three additional activities are required for this project: MileToMeter, CelsiusToFahrenheit, and FootToYard, you need to define these activities in the manifest file, which is shown in Listing 4-9.

LISTING 4-9: Manifest File to Create Multiple Activities (MultipleActivities\AndroidManifest.xml)

```xml
<?xml version="1.0" encoding="utf-8"?>
<manifest xmlns:android="http://schemas.android.com/apk/res/android"
 package="com.nfclab.multipleactivities"
 android:versionCode="1"
 android:versionName="1.0"
>

<uses-sdk android:minSdkVersion="4" />

<application
 android:icon="@drawable/ic_launcher"
 android:label="@string/app_name"
>

<activity
 android:name=".MultipleActivities"
 android:label="@string/app_name"
>
    <intent-filter>
       <action android:name="android.intent.action.MAIN" />
       <category android:name="android.intent.category.LAUNCHER" />
    </intent-filter>
</activity>

<activity
    android:name="com.nfclab.multipleactivities.MileToMeter">
</activity>

<activity
    android:name="com.nfclab.multipleactivities.CelsiusToFahrenheit">
</activity>

<activity
    android:name="com.nfclab.multipleactivities.FootToYard">
</activity>

</application>
</manifest>
```

Step 5: Creating a New Layout

To create a new layout for another activity in your project, do the following:

1. Right-click the layout folder of the current project in the Package Explorer and then click New ⇨ Other.

2. Select Android XML File from the Android menu and then click Next.

3. Name your layout file and then click Finish.

You should see that the new XML file is added to your project's layout folder. You add one new layout for each activity using the same methodology defined previously. Alternatively, you may also use one single layout for all three activities. The layout files are the same as those in the previous example, and you can see them by downloading the project's source code at `www.wrox.com/ remtitle.cgi?isbn=1118380096`.

Step 6: Creating a New Activity

To create a new activity in your project, do the following:

1. Right-click your project name in Package Explorer and then click New ⇨ Class. A new window opens (see Figure 4-18).

2. Give a name to the class, and be sure to follow Java naming conventions (such as **MileToMeter**).

3. Click Browse near Superclass and type **activity** in the text box (see Figure 4-19).

4. Select `Activity-android.app` from the list and click OK.

5. Click Finish to complete the creation of the activity.

A new class file is added to your project's `src` folder. You should implement the following classes to complete your project: `MileToMeter`, `CelsiusToFahrenheit`, and `FootToYard`.

All three activities are nearly the same as those in the previous example, so we do not describe them again. You can see them by downloading the project's source code at `www.wrox.com/remtitle. cgi?isbn=1118380096`

FIGURE 4-18

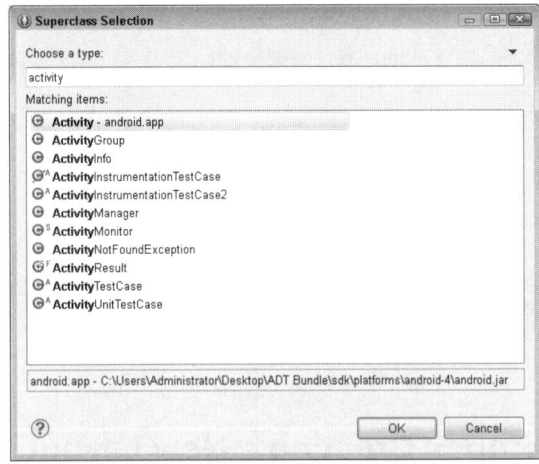

FIGURE 4-19

USING MENU ITEMS

Menus are an important part and the most commonly used user interface component in many applications. To create a consistent user experience, you should use menu APIs to present user actions and other options in your activities. Currently, Android provides a standard XML format to define menu items for all menu types. You should define a menu and all its items in an XML menu resource in your activity's code so that you can load it as a menu object in your activity. It is beneficial to use a menu as a resource; this way, you can more easily visualize the menu structure in XML and separate the content for the menu from the application's code. Furthermore, using a menu this way enables you to perform menu configuration easily for different platform versions, screen sizes, and so on.

Three types of menus are defined in Android:

➤ **Options Menu and Action Bar:** The options menu is the primary collection of menu items for an activity. It's the place where you should place actions that have a global impact on the app. Since Android 3.0, the Menu button is deprecated and the options menu is presented by the action bar.

➤ **Context Menu and Contextual Action Mode:** A context menu is a floating menu that appears when the user presses an element on the screen and holds for a second. It provides actions that affect the selected content or context frame.

➤ **Pop-up Menu:** This type of menu displays a list of items in a vertical list that is anchored to the view that invoked the menu. A pop-up menu is good for providing an overflow of actions that relate to specific content or to provide options for a second part of a command.

In the following example, you build an application with an options menu and an image displayer. We do not describe the other menu types in this chapter, and we leave it to you to find more information. Remember that you can use `http://developer.android.com/` for this purpose.

> **NOTE** *For more information on menus, refer to* `http://developer.android.com/guide/topics/ui/menus.html`.

The main screen of the application is shown in Figure 4-20. In the main screen, an image is displayed. When you press the menu button, the menu is displayed with two options: Card Emulation Mode and Peer to Peer Mode (see Figure 4-21).

FIGURE 4-20

FIGURE 4-21

Creating a Menu XML File

To create the menu file, perform the following steps:

1. Create a `menu` folder in your project's `res` folder, if you do not already have one.

2. Create an XML file in your project's `res/menu` folder.

3. Add the code shown in Listing 4-10 to your XML menu file.

LISTING 4-10: Layout Resource to Create Menu Items (ImageDisplayerwithMenu\res\menu\menu.xml)

```xml
<?xml version="1.0" encoding="utf-8"?>
<menu xmlns:android="http://schemas.android.com/apk/res/android">
    <item
      android:id="@+id/ceMode"
      android:title="Card Emulation Mode"/>
    <item
      android:id="@+id/p2pMode"
      android:title="Peer to Peer Mode"/>
</menu>
```

In this listing, `<menu>` defines a menu, which is a container for menu items. A `<menu>` element must be the root node for the file and can hold one or more `<item>` and `<group>` elements. The `<item>` element creates a `MenuItem`, which represents a single item in a menu. This element may contain a nested `<menu>` element used to create a submenu.

Another optional item named `<group>` is an invisible container for `<item>` elements. It allows you to categorize menu items so they share properties such as active state and visibility.

In the example, `android:id` and `android:title` attributes are defined for menu items:

➤ `android:id` is a resource ID that's unique to the item, which allows the application to recognize the item when the user selects it.

➤ `android:title` is a reference to a string to use as the item's title.

Additionally, the example contains two other important attributes: `android:icon` and `android:showAsAction`:

➤ `android:icon` is a reference to a drawable to use as the item's icon.

➤ `android:showAsAction` specifies when and how this item should appear as an action item in the action bar.

Layout

Actually, layout files do not need to be modified based on the created menus. However, in this example, you use `ImageView` to load images onto the screen and should modify the layout file to enable this capability. The layout file contains two `TextView` objects and an `ImageView`. In Listing 4-11, `ImageView` has two different attributes: `android:src` and `android:contentDescription`. The `android:src` attribute sets a drawable as the content of the `ImageView`. It may reference another resource or a color. In this example, it references an image file named `cemode`. You need to add the image file with the extension.`png`, `.jpg`, or `.gif` to the `res/drawable` folder in the project hierarchy. The other attribute, `android:contentDescription`, defines the text describing the content of the view.

LISTING 4-11: Layout Resource to Create ImageView (ImageDisplayerwithMenu\res\layout\main.xml)

```xml
<?xml version="1.0" encoding="utf-8"?>
<LinearLayout xmlns:android="http://schemas.android.com/apk/res/android"
 android:layout_width="fill_parent"
 android:layout_height="fill_parent"
 android:orientation="vertical"
>

<TextView
 android:id="@+id/myHeaderText"
 android:layout_width="fill_parent"
 android:layout_height="wrap_content"
 android:text="@string/headerText"
 android:textStyle="italic"
 android:gravity="center"
/>

<ImageView
 android:id="@+id/NFCImage"
 android:layout_width="wrap_content"
 android:layout_height="wrap_content"
 android:src="@drawable/cemode"
 android:layout_gravity="center"
 android:contentDescription="@string/imageDesc"
/>

<TextView
 android:id="@+id/myinfoText"
 android:layout_width="fill_parent"
 android:layout_height="wrap_content"
 android:text="@string/infoText"
 android:gravity="center"
/>

</LinearLayout>
```

Code

To create the menu in the source code, define the `onCreateOptionsMenu` method as shown in Listing 4-12. This method is called only once, when the options menu is displayed for the first time. When you use `MenuInflater`, the menu is instantiated, and the `menu.xml` file inside `res/menu` folder is loaded. When an item from the menu is selected, `onOptionsItemSelected` is called, and the method receives the selected `MenuItem` as a parameter. In the `onOptionsItemSelected` method, the `changeImage` method is invoked to change the image file, as shown in Listing 4-12. In the `changeImage` method, the selected menu item's ID is compared with all the menu items' IDs, and when the selected ID is found, the image and the header text are changed. Also note that image is an ImageView and a header is a TextView created within the constructer and instantiated in the onCreate method.

LISTING 4-12: Layout Resource to Create ImageView (ImageDisplayerwithMenu\src\com\nfclab\ imagedisplayer\ImageDisplayerwithMenu.java)

```java
public class ImageDisplayerwithMenu extends Activity {

    private ImageView image;
    private TextView header;

    @Override
    public void onCreate(Bundle savedInstanceState) {
        super.onCreate(savedInstanceState);
        setContentView(R.layout.main);
        image = (ImageView) findViewById(R.id.NFCImage);
        header = (TextView) findViewById(R.id.myHeaderText);
    }

    public void changeImage(int id) {
        if(id == R.id.ceMode){
            image.setImageResource(R.drawable.cemode);
            header.setText("Displaying Card Emulation Mode Image");
        } else if(id == R.id.p2pMode){
            image.setImageResource(R.drawable.p2pmode);
            header.setText("Displaying Peer to Peer Mode Image");
        }
    }

    @Override
    public boolean onCreateOptionsMenu(Menu menu) {
        MenuInflater menuInflater = getMenuInflater();
        menuInflater.inflate(R.menu.menu, menu);
        return true;
    }

    @Override
    public boolean onOptionsItemSelected(MenuItem item) {
        changeImage(item.getItemId());
        return true;
    }
}
```

SUMMARY

When an Android project is created in Eclipse, Eclipse automatically adds the necessary components and displays them in the Package Explorer. These components are the src folder for all the Java classes; the gen folder to hold the automatically generated files; the Android SDK version; the assets folder to put some files and later retrieve them; the res folder to hold resources, which includes the drawable, layout, and values folders by default; and AndroidManifest.xml.

You can run Android applications in two ways: on emulators and on mobile phones. In order to run on an emulator, you need to create an AVD. In order to run on a mobile phone, you need to connect the mobile to your computer and make the related configuration settings in the Android operating system.

There are different layout options that can be used to organize the screen. These are `LinearLayout`, `AbsoluteLayout`, `AdapterView`, `GridLayout`, and `RelativeLayout`. Also, you have two options to create layouts. The first option is to define them in XML files, and the other option is to define them in the program code. Using XML files for organizing layouts is the preferred option to create layouts instead of organizing them in program code.

In your application, you should externalize the application resources such as images, colors, and strings from your source code so that you can manage them independently. When you define your application resources, you can access them using resource IDs that are generated in the project's `R` class.

An event listener is an interface and can be registered to an item. When the registered item is triggered by the user, a related method is called by Android to perform the required actions.

A dialog is a small window that appears in front of an activity and gains focus for user input.

Activities are managed in a stack in Android. When an activity starts, it moves to the front of the activity stack and becomes visible. The other running activities still run if you do not quit them, but they stay in lower levels in the stack and thus become invisible. An activity mainly has four important states: active, paused, stopped, and destroyed. The `onCreate()`, `onStart()`, `onRestart()`, `onResume()`, `onPause()`, `onStop()`, and `onDestroy()` methods are used to change the states of the activities.

5

NFC Programming: Reader/Writer Mode

In this chapter, reader/writer mode application programming is demonstrated. The tag intent dispatch system, tag foreground dispatch system, and Android Application Records (AARs) are described in detail. The chapter mainly focuses on reading and writing NDEF messages from and to NFC tags. In order to write an NDEF message to a tag, at least one NDEF record needs to be created to form an NDEF message. The NDEF records are formatted with one of the Type Name Formats defined by NFC Forum. For example, if you wish to create an NDEF record containing a website URI, you may use `TNF_ABSOLUTE_URI` or `TNF_WELL_KNOWN` with `RTD_URI` type records. So this chapter also explains how to create required NDEF records and how to read them.

The chapter starts by describing the NFC APIs in the Android platform, and then continues with the tag dispatch system, NFC properties in the Android manifest file, and NFC intents. Then, writing data to tags and reading data from tags are described in detail. Finally, AARs and performing I/O operations on different tag types are described.

NFC APIS IN ANDROID

Currently, there are two packages for NFC application development in the Android platform. The first is the main package that you will use, `android.nfc`, which includes necessary classes to enable applications to read and write NDEF messages in/to NFC tags.

The second is the `android.nfc.tech` package, which includes necessary classes to provide access to different tag technologies such as MIFARE Classic, NfcA, NfcV, and so on. This package also provides input and output operations on these tags in raw bytes.

NFC-related APIs in Android are introduced to users starting from API level 9 including the `android.nfc` package. Most of the classes and methods in this package are introduced in API level 10 and more are added in API level 14 and 16. For example, in order to work with NDEF-formatted tags, you should at least use API level 10, since `ACTION_NDEF_DISCOVERED` constants to handle NDEF-formatted tags are introduced in API level 10.

On the other hand, the `android.nfc.tech` package is introduced in API level 10 and a few methods are added to the package in API level 14. The details of the packages, including the API levels, descriptions, and parameters of each class, method, and constant, are given in Appendix B.

android.nfc package

The classes and methods in the `android.nfc` package allow NFC-enabled mobile phones to read and write NDEF messages from and to supported tags. This package also enables data exchange with other NFC-enabled mobile phones. There are six classes in this package to provide these functionalities (see also Table 5-1). The first is the `Tag` class, which represents the discovered NFC tag. The second is the `NfcAdapter` class, which represents the NFC adapter of the mobile phone. The `NfcManager` class is also related to the `NfcAdapter` class and is used to obtain an instance of the NFC adapter. The two other important classes are `NdefMessage` and `NdefRecord`, which represent an NDEF message and an NDEF record, respectively. The last class is the `NFCEvent`, which wraps information associated with an NFC event.

TABLE 5-1: Classes in the android.nfc Package

CLASS NAME	DESCRIPTION
Tag	Represents the discovered NFC tag
NfcAdapter	Represents the NFC adapter of the mobile phone
NfcManager	Obtains an instance of the NFC adapter
NdefMessage	Represents an NDEF message
NdefRecord	Represents an NDEF record
NfcEvent	Wraps information associated with an NFC event

android.nfc.tech package

When a mobile phone scans a tag, the tag may not be compatible with the NDEF format. In this situation, the applications can access this tag's data in raw bytes. However, this access and I/O operation in raw bytes should be different in different tag types. This package includes the classes required by different tag types in order to perform I/O operations on them. Some of the classes are IsoDep, MifareClassic, MifareUltralight, and NfcV. The package also includes the TagTechnology interface, which is needed in order to obtain the tag and connect to it. The details of this package will be described in the next sections, together with examples.

TAG INTENT DISPATCH SYSTEM VS. FOREGROUND DISPATCH SYSTEM

The tag intent dispatch system is used to launch applications when the predefined tags or NDEF data are identified in tags. In short, you scan to an NFC tag, and if any application is registered to handle the tag, then the registered application launches. If more than one application is registered to handle the tag, a pop-up to select the application (Activity Chooser) is displayed. For example, if your application is coded with the tag intent dispatch system to detect any MIME type data and a tag is discovered that contains the text/plain data type your application is launched. However, when your application is in active state and the same tag is discovered, Android will again run your application, or if there is more than one application that can handle the text/plain data type, it will ask you to select one of the applications that can handle the tags by displaying Activity Chooser. However, in modified versions of Android (as in some OEM implementations), this can be changed so that if one application is in active state, that application can be used.

The foreground dispatch system, on the other hand, is designed to handle tags when the application is running. When you run an application that uses the foreground dispatch system and is registered to detect tags with any MIME type, and a tag is discovered that contains the `text/plain` data type your application will handle the tag. Moreover, Android will not display Activity Chooser, even though there are other applications that can handle the tag.

The difference in the coding is that the tag intent dispatch system registers the tag types and NDEF data that the application can handle in the application's manifest file using intent filters, whereas the foreground dispatch system registers inside the activity.

NFC TAG INTENT DISPATCH SYSTEM

When an NFC tag is scanned, the desired action is to launch the corresponding application automatically. Then the application should perform the required activities with the data transferred from the tag. In this way, the usability of the NFC technology will be high.

In Android, when NFC is not disabled from settings, the mobile phone always looks for NFC tags to discover. When an NFC tag is discovered in proximity, the type and payload data in the tag will be encapsulated to intent, and the tag intent dispatch system in Android will run the corresponding application that can handle the tag. For this reason, applications can register the type of the data that they can handle. Also, in order to handle only the tags that your activity looks for, you should register only the data that your application can handle. If multiple applications are registered to handle the same type of data, the Activity Chooser will be displayed for the user to select one of the applications.

The NFC tag dispatch system in Android works in the following way:

1. It parses the NFC tag and tries to figure out the type of the payload data inside the tag (for example, MIME type or URI).

2. It encapsulates the type and the payload in an intent.

3. If an installed activity is registered to handle this payload (based on the type of the payload), related activity is started.

How NFC Tags Are Dispatched to Applications

In a typical Android NFC application that scans an NFC tag, there can be three different options:

➤ The scanned tag contains NDEF payload that *can* be mapped to a MIME type or URI

➤ The scanned tag contains NDEF payload that *cannot* be mapped to a MIME type or URI

➤ The scanned tag does not contain NDEF payload but is of a known tag technology

Based on these options, three different intents are developed: `ACTION_NDEF_DISCOVERED`, `ACTION_TECH_DISCOVERED`, and `ACTION_TAG_DISCOVERED`.

ACTION_NDEF_DISCOVERED

This intent is the highest priority among the others. When a tag contains NDEF data, the tag dispatch system tries to run an activity with this intent. If at least one application is registered to handle the type of the discovered NDEF payload, Android system runs the corresponding application.

ACTION_TECH_DISCOVERED

This intent has the second highest priority. If the discovered payload is NDEF and no application is registered to handle the discovered NDEF payload type, then intent with `ACTION_TECH_DISCOVERED` is created, and the tag dispatch system tries to start an activity with this intent.

If the type of the payload in the discovered tag is not NDEF and the type of the tag can be recognized by Android system, then the tag intent dispatch system creates `ACTION_TECH_DISCOVERED` intent.

ACTION_TAG_DISCOVERED

This intent has the lowest priority. If no activities in the mobile device can handle the corresponding `ACTION_NDEF_DISCOVERED` or `ACTION_TECH_DISCOVERED` intents, `ACTION_TAG_DISCOVERED` is created.

When the device scans the tag, the tag dispatch system prepares an intent that encapsulates the NFC tag and its identifying information. Then the applications that can handle the prepared intent are found. If only one application is found that can handle the intent, the application is started. If more than one application is found, the Activity Chooser is displayed so that the user can select the one that they want.

The tag dispatch system works as follows (which is also shown in Figure 5-1):

➤ If the payload contains NDEF data, do the following:

   1. Try to start an activity with the `ACTION_NDEF_DISCOVERED` intent.

   2. If no activities filter for the discovered NDEF data, try to start an activity with the `ACTION_TECH_DISCOVERED` intent.

➤ If the payload does not contain NDEF data but is of a known tag technology, do the following:

   1. Try to start an activity with the `ACTION_TECH_DISCOVERED` intent.

   2. If no activities filter for that intent, try to start an activity with the `ACTION_TAG_DISCOVERED` intent.

➤ If there's no applications filter for the encapsulated intent, do nothing.

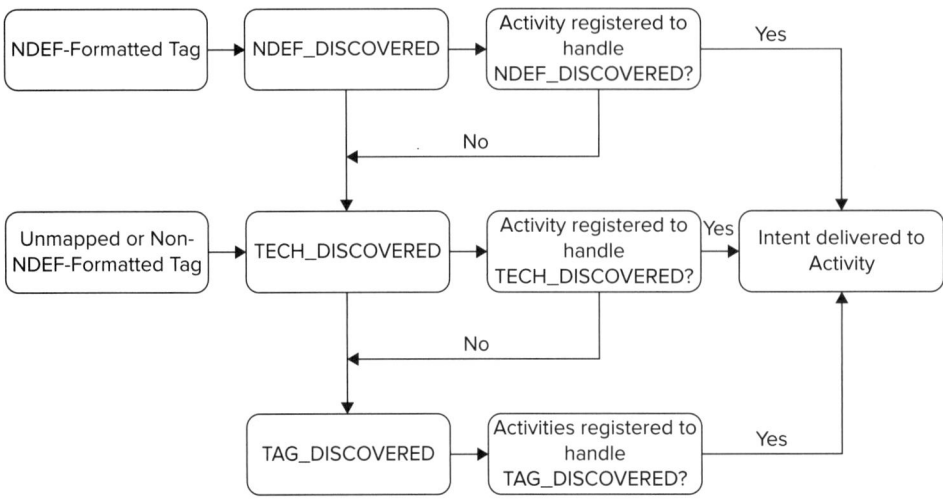

FIGURE 5-1

How Android Handles NDEF-Formatted Tags

As described in Chapter 2, "NFC Essentials for Application Developers," NDEF data is encapsulated inside an NDEF message that contains one or more NDEF records. When a tag is discovered, the first thing that an Android-powered device does is to create intent of ACTION_NDEF_DISCOVERED. It can create this intent if the device can determine the type of the data based on the first NDEF record. For example, if the type of the data is a MIME type or a URI, the device creates the ACTION_NDEF_DISCOVERED intent. In order for an Android device to determine the type of the data, the Type Name Format (TNF) of the data should be one of the formats given in Table 5-2.

TABLE 5-2: Supported TNFs and their Mappings

TNF	MAPPING
TNF_ABSOLUTE_URI	Absolute URI based on the type field
TNF_EMPTY	Falls back to ACTION_TECH_DISCOVERED intent
TNF_EXTERNAL_TYPE	URI based on the Uniform Resource Name (URN) in the type field; the URI is saved to the tag in \<domain_name\>:\<service_name\> form; Android converts it automatically to the form of vnd. android .nfc://ext/\<domain_name\>:\<service_name\>
TNF_MIME_MEDIA	Data with a MIME type based on the type field
TNF_UNCHANGED	Falls back to ACTION_TECH_DISCOVERED intent
TNF_UNKNOWN	Falls back to ACTION_TECH_DISCOVERED intent
TNF_WELL_KNOWN	Described in Table 5-3

TABLE 5-3: Supported RTDs for TNF_WELL_KNOWN and Their Mappings

RECORD TYPE DEFINITION (RTD)	MAPPING
RTD_ALTERNATIVE_CARRIER	Falls back to `ACTION_TECH_DISCOVERED` intent
RTD_HANDOVER_CARRIER	Falls back to `ACTION_TECH_DISCOVERED` intent
RTD_HANDOVER_REQUEST	Falls back to `ACTION_TECH_DISCOVERED` intent
RTD_HANDOVER_SELECT	Falls back to `ACTION_TECH_DISCOVERED` intent
RTD_SMART_POSTER	URI based on parsing the payload
RTD_TEXT	Data with a MIME type of text/plain
RTD_URI	URI based on payload

If the discovered tag contains NDEF data but has a data format other than the ones listed in Table 5-2, then the device cannot create the intent of `ACTION_NDEF_DISCOVERED`. This time the device creates `ACTION_TECH_DISCOVERED` intent instead.

Some of the TNFs cannot be mapped to a MIME type or URI, and when these formats are discovered, the tag intent dispatch system creates `ACTION_TECH_DISCOVERED` instead of `ACTION_NDEF_DISCOVERED` intent. For example, if the TNF of the record is `TNF_EMPTY`, then it falls back to `ACTION_TECH_DISCOVERED` intent.

NFC PROPERTIES IN THE ANDROID MANIFEST FILE

In order to enable NFC technology in an application, you should give the required permissions to the application to use the NFC hardware. Giving this permission will enable your application to handle the intents and use NFC hardware.

First, you should enable NFC by declaring the following line in the Android manifest file:

```
<uses-permission android:name="android.permission.NFC" />
```

For the minimum SDK version that your application can support, we suggest that you use at least API level 10, because most of the functions are provided in API level 10. In API level 9, only `ACTION_TAG_DISCOVERED` intent is introduced, and other intents such as `ACTION_NDEF_DISCOVERED` and `ACTION_TECH_DISCOVERED` are introduced in API level 10. Foreground NDEF pushing, which will be described later, is also introduced in API level 10.

On the other hand, in API level 14, extra methods and Android Beam are introduced, which enable two devices to share NDEF messages. The details of each method, including their API levels, are given in Appendix B.

You need to give interest in the manifest file. You should modify the `uses-sdk` element and at least use API level 10, as shown here:

```
<uses-sdk android:minSdkVersion="10"/>
```

When you publish your NFC application in the Android market, all users will be able to download and install your application. If you wish only the users who have a mobile with the required NFC hardware to see and download your application, you should add the following line to the Android manifest file:

```
<uses-feature android:name="android.hardware.nfc" android:required="true" />
```

You may also omit the `uses-feature` element and allow all users to download your application from the Android market. This time, you should check if the mobile phone has an NFC antenna in your application. If the mobile phone doesn't have an NFC antenna, then you should warn the user that the mobile device doesn't have the required NFC functionality.

FILTERING NFC INTENTS

You generally want your application to start when a tag that your application can handle is scanned. Inside three different intents, you want `ACTION_NDEF_DISCOVERED` intent, since it has the highest priority. Let's say you deploy a tag that contains `RTD_URI` data. You need to register your application to handle `RTD_URI` with `ACTION_NDEF_DISCOVERED` intent. If you register your application with `ACTION_TECH_DISCOVERED` intent, then your application will probably not start automatically when the tag is scanned, since `ACTION_TECH_DISCOVERED` intent is a fallback for `ACTION_NDEF_DISCOVERED` and there will probably be at least one application installed on the mobile phone that handles `RTD_URI` with `ACTION_NDEF_DISCOVERED` intent.

On the other hand, you should generally never use `ACTION_TAG_DISCOVERED` intent, since it is too general to filter and your application will be able to filter only the tags that cannot be handled by other applications.

In most of the situations, especially when you control and deploy NFC tags, you should use NDEF data in your tags, and you should always use `ACTION_NDEF_DISCOVERED` intents in your applications. In the following subsections, filtering for three different intents is described.

> **NOTE** *Android intent filtering is always case sensitive.*

ACTION_NDEF_DISCOVERED

In order to filter `ACTION_NDEF_DISCOVERED` intents, the intent should be defined along with the type of the data that need to be filtered. The type of the data may be the TNFs that are described in previous sections, such as `RTD_TEXT`, `RTD_URI`, `TNF_EXTERNAL_TYPE`, and so on.

A general declaration to filter `ACTION_NDEF_DISCOVERED` is shown here:

```
<activity>
...
   <intent-filter>
   <action android:name="android.nfc.action.NDEF_DISCOVERED" />
      <category android:name="android.intent.category.DEFAULT" />
```

```
        <data ...
            ... />
    </intent-filter>
...
</activity>
```

The `action` element is set to `android.nfc.action.NDEF_DISCOVERED`, because the action specified in the intent filter must match one of the actions listed in the filter. This means that when a tag that has NDEF-formatted data in it, the action will match the intent filter. The `category` element is set to `android.intent.category.DEFAULT`, because the activities that need to receive implicit intents need to include `android.intent.category.DEFAULT` in the intent filter. The `data` element needs to change based on the TNFs that are to be filtered. The detailed manifests to filter each TNF are given in the following subsections.

Declaring an Intent Filter for TNF_WELL_KNOWN with RTD_URI

In order to filter `RTD_URI` records, the intent filter should be declared as follows:

```
<activity>
...
    <intent-filter>
        <action android:name="android.nfc.action.NDEF_DISCOVERED" />
            <category android:name="android.intent.category.DEFAULT" />
                <data android:scheme="http"
                    android:host="nfclab.com"
                    android:pathPrefix="" />
    </intent-filter>
...
</activity>
```

As you can see in the code, the data element includes attributes to filter `RTD_URI`. `android:scheme` filters for the scheme of the URI (for example, http, ftp, etc.) and `android:host` filters for the host of the domain (for example, `nfclab.com`). When a tag's first record is the NDEF `RTD_URI` record of `http://nfclab.com`, the intent filter will filter the intent and launch the corresponding activity. If the NDEF record inside the tag has a `www.` prefix, then the host should be `www.nfclab.com`.

Declaring an Intent Filter for TNF_WELL_KNOWN with RTD_TEXT

In order to filter `RTD_TEXT` records, the intent filter should be declared as follows:

```
<activity>
...
    <intent-filter>
        <action android:name="android.nfc.action.NDEF_DISCOVERED" />
            <category android:name="android.intent.category.DEFAULT" />
                <data android:mimeType="text/plain" />
    </intent-filter>
...
</activity>
```

As shown in the code, the data element has an `android:mimeType` attribute with its value set to `text/plain`. When the first NDEF record in the NDEF message has `TNF_WELL_KNOWN` TNF and its RTD is `RTD_TEXT`, the given `intent-filter` will filter the intent and launch the corresponding activity.

Declaring an Intent Filter for TNF_ABSOLUTE_URI

In order to filter `TNF_ABSOLUTE_URI` records, the intent filter should be declared as follows:

```
<activity>
...
    <intent-filter>
        <action android:name="android.nfc.action.NDEF_DISCOVERED" />
            <category android:name="android.intent.category.DEFAULT" />
                <data android:scheme="http"
                    android:host="nfclab.com"
                    android:pathPrefix="/index.html" />
    </intent-filter>
...
</activity>
```

As you can see in the code, the `intent-filter` is very similar to the intent filter of `RTD_URI`. The only difference is that the `android:pathPrefix` attribute is not empty and its value is set to the filename and the extension. When a tag has the first record of `TNF_ABSOLUTE_URI` of `http://nfclab.com/index.html`, the `intent-filter` will filter the intent and launch the corresponding activity.

Declaring an Intent Filter for TNF_MIME_MEDIA

In order to filter `TNF_MIME_MEDIA` records, the intent filter should be declared as shown here:

```
<activity>
...
    <intent-filter>
        <action android:name="android.nfc.action.NDEF_DISCOVERED" />
            <category android:name="android.intent.category.DEFAULT" />
                <data android:mimeType="application/NFCLabApp" />
    </intent-filter>
...
</activity>
```

As shown in the code, the data element has an `android:mimeType` attribute with its value set to `application/NFCLabApp`. When a tag has the first record of `TNF_MIME_MEDIA`, the intent filter will filter the intent and launch the corresponding activity.

The following code gives another intent filter for a known `TNF_MIME_MEDIA` record: `text/x-vCard`, which is a standard file format for electronic business cards. When a business card with `text/x-vCard` is discovered inside the first NDEF record, the corresponding activity will be launched:

```
<activity>
...
    <intent-filter>
        <action android:name="android.nfc.action.NDEF_DISCOVERED" />
            <category android:name="android.intent.category.DEFAULT" />
```

```
                 <data android:mimeType="text/x-vCard" />
        </intent-filter>
    ...
    </activity>
```

If you define */* value in the android:mimeType, then your application will filter for all
TNF_MIME_MEDIA records.

Declaring an Intent Filter for TNF_EXTERNAL_TYPE

URNs for TNF_EXTERNAL_TYPE have the format of urn:nfc:ext:<domain_name>:<service_
name>. However the NDEF record in the tag should not store the urn:nfc:ext: part. So, when
you're creating a TNF_EXTERNAL_TYPE record, you should only provide the domain name and service
name of the record. When Android processes an NDEF record that contains TNF_EXTERNAL_TYPE, it
converts the URN of urn:nfc:ext:<domain_name>:<service_name> into vnd.android.nfc://
ext/<domain_name>:<service_name>. So, in order to filter TNF_EXTERNAL_TYPE records, the intent
filter should be formed as shown here:

```
<activity>
...
    <intent-filter>
        <action android:name="android.nfc.action.NDEF_DISCOVERED" />
            <category android:name="android.intent.category.DEFAULT" />
                <data android:scheme="vnd.android.nfc"
                    android:host="ext"
                    android:pathPrefix="/nfclab.com:customService" />
    </intent-filter>
...
</activity>
```

As shown in the code, the data element includes three attributes: android:scheme, android:host,
and android:pathPrefix. The android:scheme and android:host attributes should stay the
same, and the android:pathPrefix attribute should be personalized. This is where the self-
allocation of the record is performed. When a tag has the first record of TNF_EXTERNAL_TYPE
with nfclab.com:customService, the intent-filter will filter the intent and launch the
corresponding activity.

If you define *:* value in the android:pathPrefix, then your application will filter for all
TNF_EXTERNAL_TYPE records.

ACTION_TECH_DISCOVERED

Remember that ACTION_TECH_DISCOVERED is used when NDEF data cannot be mapped to a MIME
type or a URI, or when the tag does not contain NDEF data. In order to filter for ACTION_TECH_
DISCOVERED intents, the AndroidManifest.xml file should be described as follows:

```
<activity>
...
    <intent-filter>
        <action android:name="android.nfc.action.TECH_DISCOVERED" />
```

```
        </intent-filter>
        <meta-data android:name="android.nfc.action.TECH_DISCOVERED"
            android:resource="@xml/nfc_tech_list" />
...
    </activity>
```

The `android:resource` attribute in the `meta-data` element points to an XML file that consists of the tag technologies searched for and described in the following paragraphs.

The `android.nfc.tech` package consists of 10 classes: `Ndef`, `NdefFormatable`, `IsoDep`, `Mifare Classic`, `MifareUltralight`, `NfcA`, `NfcB`, `NfcBarcode`, `NfcF`, and `NfcV`. Each class represents a different tag technology. These tag technologies are used for `ACTION_TECH_DISCOVERED` intent.

> **NOTE** *For more information about the* `android.nfc.tech` *package, refer to Appendix B.*

When your application filters for `ACTION_TECH_DISCOVERED` intent, the intent created by the tag dispatch system searches for the available tag technologies defined by the application. The application needs to define these available tag technologies within a tech-list set inside an XML file (the filename does not matter) in the `/res/xml` folder. The following code gives an example tech-list set containing all the technologies:

```
<resources xmlns:xliff="urn:oasis:names:tc:xliff:document:1.2">
    <tech-list>
        <tech>android.nfc.tech.IsoDep</tech>
        <tech>android.nfc.tech.NfcA</tech>
        <tech>android.nfc.tech.NfcB</tech>
        <tech>android.nfc.tech.NfcF</tech>
        <tech>android.nfc.tech.NfcV</tech>
        <tech>android.nfc.tech.Ndef</tech>
        <tech>android.nfc.tech.NdefFormatable</tech>
        <tech>android.nfc.tech.MifareClassic</tech>
        <tech>android.nfc.tech.MifareUltralight</tech>
<tech>android.nfc.tech.NfcBarcode</tech>
    </tech-list>
</resources>
```

When a tag is scanned and the `ACTION_TECH_DISCOVERED` intent is filtered, the defined tech-list should be a subset of the scanned tag's supported technologies. For example, if the scanned tag's supported technologies are `IsoDep`, `NfcA`, and `MifareClassic`, the defined tech-list should specify one, two, or three of these technologies but nothing else.

The technologies supported by the tag can be obtained using the `getTechList()` method, which will be described later.

More than one tech-list can be defined in a tech-list XML file, and if one of the tech-lists is a subset of the tag's technologies, it is considered a match.

For example, consider the following tech-list XML file:

```
<resources xmlns:xliff="urn:oasis:names:tc:xliff:document:1.2">
   <tech-list>
      <tech>android.nfc.tech.NfcB</tech>
      <tech>android.nfc.tech.Ndef</tech>
   </tech-list>
</resources>

<resources xmlns:xliff="urn:oasis:names:tc:xliff:document:1.2">
   <tech-list>
      <tech>android.nfc.tech.MifareClassic</tech>
      <tech>android.nfc.tech.NfcA</tech>
   </tech-list>
</resources>
```

Consider also that the scanned tag supports `MifareClassic`, `NdefFormatable`, and `NfcA` technologies. Then there will be a match, because the second tech-list in the XML file contains `MifareClassic` and `NfcA` and it is a subset of the tag's technologies.

ACTION_TAG_DISCOVERED

`ACTION_TAG_DISCOVERED` intent is started if no activities handle the `ACTION_NDEF_DISCOVERED` or `ACTION_TECH_DISCOVERED` intents; however, your application should be well designed so that you should not need to use this intent. The required XML code for the `AndroidManifest.xml` file to filter for `ACTION_TAG_DISCOVERED` intent is as follows:

```
<activity>
...
   <intent-filter>
      <action android:name="android.nfc.action.TAG_DISCOVERED"/>
   </intent-filter>
...
</activity>
```

CHECKING THE NFC ADAPTER

The `NFCAdapter` class in the `android.nfc` package represents the NFC adapter for the local machine. Checking if the `NFCAdapter` is available is the first thing to do in an NFC application. Here is the code to create the necessary `NfcAdapter`:

```
private NfcAdapter myNfcAdapter;
private TextView myText;

@Override
public void onCreate(Bundle savedInstanceState) {
   super.onCreate(savedInstanceState);
   setContentView(R.layout.main);
   myText= (TextView) findViewById(R.id.myText);
   myNfcAdapter = NfcAdapter.getDefaultAdapter(this);
   if (myNfcAdapter == null)
      myText.setText("NFC is not available for the device!!!");
   else
      myText.setText("NFC is available for the device");
}
```

As you can see from the code, a new `NFCAdapter` named `myNfcAdapter` is created. `getDefaultAdapter(Context)` is used to get the default NFC adapter for the device. The method returns the default NFC adapter or returns null if the NFC adapter does not exist. In this context, the `NfcAdapter.getDefaultAdapter(this)` method returns null or the current NFC adapter to the `myNfcAdapter` object. Finally, informative text is displayed on the screen to notify the user if the NFC adapter exists.

TAG WRITING

When the application discovers a tag, it starts an activity that is defined in the `AndroidManifest.xml` file. Inside the activity, you may perform different operations based on the discovered tag. You may start an intent with `ACTION_NDEF_DISCOVERED` (when the tag with NDEF payload is discovered), with `ACTION_TECH_DISCOVERED` (when non-NDEF data is discovered or NDEF data cannot be mapped to a TNF and the tag technology is identified by Android), or with `ACTION_TAG_DISCOVERED` (when a tag is discovered).

In the following code, the basic write operation to a tag is shown, in which the discovered tag has NDEF payload:

```
if (NfcAdapter.ACTION_NDEF_DISCOVERED.equals(getIntent().getAction())){
    Tag detectedTag = getIntent().getParcelableExtra(NfcAdapter.EXTRA_TAG);

    // PREPARE THE NDEF MESSAGE

    // WRITE DATA TO TAG
}
```

First, the intent is checked if it was launched from an NFC interaction of `NDEF_DISCOVERED`:

```
if (NfcAdapter.ACTION_NDEF_DISCOVERED.equals(getIntent().getAction())){
```

Then the instance of the discovered tag is received and saved to the `detectedTag` object:

```
Tag detectedTag = getIntent().getParcelableExtra(NfcAdapter.EXTRA_TAG);
```

`Tag` represents the state of an NFC tag at the time of discovery. The `getParcelableExtra` method is used to retrieve the extended data from the intent. Moreover, `NfcAdapter.EXTRA_TAG` is used as a parameter to get the tag information.

After getting the instance of the detected tag, first the NDEF message needs to be prepared and then it should be written to the tag. These operations are described in the following sections.

Preparing NDEF Data

In order to write NDEF messages to tags, the corresponding NDEF records and NDEF message need to be prepared based on the desired data type. The data preparation needs to be made based on the NFC Forum standards. The standards are described in Chapter 2 and can also be accessed online via the NFC Forum website (http://www.nfc-forum.org).

After preparing the message, the message will be written to the tag, which is described after this section.

TNF_WELL_KNOWN with RTD_URI

In order to create a `TNF_WELL_KNOWN` record with `RTD_URI`, you need to create the URL together with its prefix. Prefixes can be, for example, `"http://www."`, `"http://"`, or `"https://"`. Each prefix has its own byte code; for example, `"http://www."` equals 1, `"http://"` equals 3, and no prefix equals 0. Available prefixes and their equivalent bytes can be seen in Appendix A. For each record, the prefixes should be included before the URI data. In the following code, a `TNF_WELL_KNOWN` record with `RTD_URI` is created with the payload data of `nfclab.com` and with the prefix of `0x01`:

```
byte[] uriField = "nfclab.com".getBytes(Charset.forName("US-ASCII"));
byte[] payload = new byte[uriField.length + 1];
payload[0] = 0x01;
System.arraycopy(uriField, 0, payload, 1, uriField.length);
NdefRecord uriRecord = new NdefRecord(
    NdefRecord.TNF_WELL_KNOWN, NdefRecord.RTD_URI, new byte[0], payload);
NdefMessage newMessage= new NdefMessage(new NdefRecord[] { uriRecord });
```

In the code, first the URI is created and saved into the `uriField` byte array by encoding it with the US-ASCII character set. Then the `payload` byte array to store the entire payload is created with the size of the URI and an additional byte to store the prefix. Since the prefix of `"http://www."` is used for this example, the first byte of the payload is set to `0x01`. Then, the URI data is added to the payload array using the `System.arraycopy` method. This method simply copies `uriField.length` elements from the array `uriField`, starting at offset 0, into the array `payload`, starting at offset 1.

At API level 14, two new methods are defined for shortly creating `RTD_URI` records. The first one is the `createUri (String)` method, and the second one is the `createUri (Uri)` method. Instead of manually creating the `RTD_URI` NDEF record, you may use one of the following codes below:

```
NdefRecord uriRecord = NdefRecord.createUri ("http://www.nfclab.com/");
NdefRecord uriRecord = NdefRecord.createUri (Uri.parse("http://www.nfclab.com/"));
```

Then, the last thing to do is to create an NDEF message by creating the NDEF record with `RTD_URI`. In order to perform this, you need to create an `NdefRecord` using the `NdefRecord` class. An `NdefRecord` that is not constructed from the raw bytes should contain the following:

➤ 3-bit TNF field, which indicates how to interpret the type field

➤ Variable length type, which describes the format of the record

➤ Variable length ID, which is the identifier of record

➤ Variable length payload, which is the actual data

The TNF for the `RTD_URI` is `TNF_WELL_KNOWN` and the record type format is `RTD_URI`. Since this NDEF record will be the first record in the NDEF message, the ID of 0 is given. Finally, the actual payload is given to the `NdefRecord` constructor as a last parameter, as shown here:

```
NdefRecord(NdefRecord.TNF_WELL_KNOWN,NdefRecord.RTD_URI,new byte[0],payload)
```

In order to form the NDEF message that will be written to the tag, you need to use the `NdefMessage` class. An NDEF message may contain one or more NDEF records. In the example, only one NDEF record is stored using the following line of code:

```
NdefMessage newMessage= new NdefMessage(new NdefRecord[] { uriRecord });
```

In order to store more than one NDEF record, the `NdefMessage` can be constructed as follows:

```
NdefMessage newMessage= new NdefMessage(
    new NdefRecord[] { uriRecord1,uriRecord2, uriRecord3 });
```

TNF_WELL_KNOWN with RTD_TEXT

As described in Chapter 2, in `RTD_TEXT` records, the first four bytes form the NDEF record header whereas the remaining part is the payload. The payload consists of the status byte (1 byte), the language code of the text (the size is gathered from the status byte), and the actual text.

In `RTD_TEXT` records, the text can be encoded in either UTF-8 or UTF-16, which is defined by the status byte in the text record. Additionally, the status byte also includes the length of the language code, which is used to identify the language code of the text. UTF-8 is identified by the bytes of `0x00` whereas UTF-16 is identified by `-0x80`. For example, if you select the encoding as UTF-8 and a language code with a length of 2 bytes; then the status byte becomes `0x02` (`0x00` plus `0x02`).

In the following code, the `RTD_TEXT` record is created in such a way that it matches with the requirements of the `RTD_TEXT` record:

```
Locale locale= new Locale("en","US");
byte[] langBytes = locale.getLanguage().getBytes(Charset.forName("US-ASCII"));
boolean encodeInUtf8=false;
Charset utfEncoding = encodeInUtf8 ? Charset.forName("UTF-8") :
    Charset.forName("UTF-16");
int utfBit = encodeInUtf8 ? 0 : (1 << 7);
char status = (char) (utfBit + langBytes.length);
String RTD_TEXT= "This is an RTD_TEXT";
byte[] textBytes = RTD_TEXT.getBytes(utfEncoding);
byte[] data = new byte[1 + langBytes.length + textBytes.length];
data[0] = (byte) status;
System.arraycopy(langBytes, 0, data, 1, langBytes.length);
System.arraycopy(textBytes, 0, data, 1 + langBytes.length, textBytes.length);
NdefRecord textRecord = new NdefRecord(NdefRecord.TNF_WELL_KNOWN,
    NdefRecord.RTD_TEXT, new byte[0], data);
NdefMessage newMessage= new NdefMessage(new NdefRecord[] { textRecord });
```

First, a new `Locale` is created with US English. (Detailed information about `Locale` can be found at: `http://developer.android.com/reference/java/util/Locale.html`.) Then a new byte array named `langBytes` is created from the `Locale`.

In this example, UTF-16 is preferred and a boolean `encodeInUtf8` value is created and initialized as `false`. Then, a charset is created named `utfEncoding` to store the charset of the UTF-8 or UTF-16. In the example, it stores the charset of UTF-16. In order to create the status byte, again the `encodeInUtf8` value is checked and the status byte is created which also includes the length of the language code.

Since there are two options for encoding, the actual text is encoded with the selected encoding. Then the byte array that will contain the actual payload data is created, named `data`. One byte for the status, the bytes required for the language code, and the bytes required for the actual text are reserved. Then the status byte, language code, and actual text are copied to the `data` array.

The last thing to do is to create the required NDEF record and save it into the NDEF message. The NDEF record is created in the same way as `RTD_URI`; the only difference is that the variable length type is defined as `RTD_TEXT`.

TNF_ABSOLUTE_URI

`TNF_ABSOLUTE_URI` indicates the absolute form of a URI that follows the `absolute-URI` rule defined by RFC 3986. The required code to create an absolute URI record is as follows:

```
byte[] uri = "http://nfclab.com/index.html".getBytes(Charset.forName("US-ASCII"));
NdefRecord uriRecord = new NdefRecord( NdefRecord.TNF_ABSOLUTE_URI,
                                       uri, new byte[0], new byte[0] );
NdefMessage newMessage= new NdefMessage(new NdefRecord[] { uriRecord });
```

As you can see in the code, first a byte array named `uri` containing the absolute URI is created by encoding it with the US-ASCII charset and is then used to create a `TNF_ABSOLUTE_URI` record by selecting the record type as `NdefRecord.TNF_ABSOLUTE_URI`.

TNF_MIME_MEDIA

The following related code is used to create a `TNF_MIME_MEDIA` record:

```
String mimeType = "application/nfclabapp";
String payload = "This is a TNF_MIME_MEDIA";
NdefRecord mimeRecord = new NdefRecord(NdefRecord.TNF_MIME_MEDIA ,
    mimeType.getBytes(), new byte[0],
    payload.getBytes(Charset.forName("US-ASCII")));
NdefMessage newMessage= new NdefMessage(new NdefRecord[] { mimeRecord });
```

As you can see in the code, the record has a MIME type of `application/nfclabapp`, which is stored in the `mimeType` string. In order to create the `NdefRecord`, the TNF to create the record is defined as `NdefRecord.TNF_MIME_MEDIA` and the variable length type of the NDEF record is defined as the bytes of the MIME type. Finally, the identifier of the record is given as 0, since it will be the first record in the `NdefMessage`, and the payload that will be written to the tag is retrieved from the `payload` string.

At API level 16, a new method is defined to create `TNF_MIME_MEDIA` in a short way. The method gets two parameters — the MIME type and the payload — as a byte array, and creates the corresponding NDEF record. You may use the method as follows:

```
NdefRecord mimeRecord = NdefRecord.createMime("application/nfclabapp",
    payload.getBytes());
```

TNF_EXTERNAL_TYPE

In order to create a `TNF_EXTERNAL_TYPE` record, you need to create the NDEF message as shown here:

```
String externalType = "nfclab.com:customService";
String payload = "texttowrite";
NdefRecord extRecord = new NdefRecord(NdefRecord.TNF_EXTERNAL_TYPE,
    externalType.getBytes(), new byte[0], payload.getBytes());
NdefMessage newMessage = new NdefMessage(new NdefRecord[] { extRecord });
```

The registered service for the record is given in the string of the `externalType` whereas the registered payload is given in the string of the `payload`. The type name format to create the record is defined as `TNF_EXTERNAL_TYPE`, and the variable length type is defined as the bytes of the registered service.

Starting from API level 16, you may use a new `createExternal ()` method, which automatically creates the `TNF_EXTERNAL_TYPE` NDEF record. The method requires three parameters as a byte array: the domain name, the domain-specific data type, and the payload.

```
NdefRecord extRecord = NdefRecord.createExternal("nfclab.com", "customService",
    payload.getBytes());
```

Writing NDEF Data to Tags

This section describes how to write NDEF data to tags. Each NDEF record should be prepared based on its type. For example, `RTD_URI` records and `TNF_MIME_MEDIA` records are to be prepared differently. Since the tag dispatch system prepares the intent from the discovered tag based on the first NDEF record, the given NDEF record examples need to be placed in the first NDEF record in order for intent filters to filter those intents.

When the data is prepared, the write operation is the same for all record types. When an NDEF message is prepared, the `writeNdefMessageToTag` method is called with two parameters, which are the prepared NDEF message and the detected tag as shown here:

```
writeNdefMessageToTag(newMessage, detectedTag);
```

The `writeNdefMessageToTag` method that writes the NDEF message to the tag is as follows:

```
boolean writeNdefMessageToTag(NdefMessage message, Tag detectedTag) {
    int size = message.toByteArray().length;
    try {
        Ndef ndef = Ndef.get(detectedTag);
        if (ndef != null) {
            ndef.connect();
            if (!ndef.isWritable()) {
                Toast.makeText(this, "Tag is read-only.", Toast.LENGTH_SHORT).show();
                return false;
            }
            if (ndef.getMaxSize() < size) {
                Toast.makeText(this, "The data cannot written to tag,
                        Tag capacity is " + ndef.getMaxSize() + " bytes, message is "
                        + size + " bytes.", Toast.LENGTH_SHORT).show();
                return false;
            }
```

```
            ndef.writeNdefMessage(message);
            ndef.close();
            Toast.makeText(this, "Message is written tag.",
                        Toast.LENGTH_SHORT).show();
            return true;
        } else {
            NdefFormatable ndefFormat = NdefFormatable.get(detectedTag);
            if (ndefFormat != null) {
                try {
                    ndefFormat.connect();
                    ndefFormat.format(message);
                    ndefFormat.close();
                    Toast.makeText(this, "The data is written to the tag ",
                                Toast.LENGTH_SHORT).show();
                    return true;
                } catch (IOException e) {
                    Toast.makeText(this, "Failed to format tag",
                                Toast.LENGTH_SHORT).show();
                    return false;
                }
            } else {
                Toast.makeText(this, "NDEF is not supported",
                            Toast.LENGTH_SHORT).show();
                return false;
            }
        }
    } catch (Exception e) {
        Toast.makeText(this, "Write opreation is failed",
            Toast.LENGTH_SHORT).show();
    }
    return false;
}
```

In Android, `Ndef` is a class in the `android.nfc.tech` package and provides access to NDEF content on a tag. Inside the method, first the instance of the `Ndef` object is acquired from the tag and saved into the `ndef` object with the following code:

```
Ndef ndef = Ndef.get(detectedTag);
```

Then the method checks the `ndef` object to see if it's null or not. If it is null it means that it is not NDEF-formatted and you cannot use `Ndef` class. If it is already NDEF-formatted, you will use `Ndef` class; otherwise, you need to use `NdefFormatable` class:

```
if (ndef != null){ … }
```

In order to enable I/O operations on the tag, the `connect()` method is applied on the `ndef` object. When the I/O operations are completed, the object should be closed with the `close()` method:

```
ndef.connect();
…
ndef.close();
```

Then the `isWritable()` method is used to find out if the tag is writable. If the tag is write-protected, the operation is cancelled:

```
if (!ndef.isWritable()) { }
```

The tag is checked if it is large enough to store the size of the NDEF message that is to be written to the tag. If the message is bigger than the tag's available size, then the operation is cancelled:

```
if (ndef.getMaxSize() < size) {}
```

Then the NDEF message is written to the tag using the following code:

```
ndef.writeNdefMessage(message);
```

The next part of the code (the first else block) is for the tags whose `ndef` object is found as null. It means that the tag is not NDEF-formatted. In this case, you need to get the instance of the tag to the `NdefFormatable` object. `NdefFormatable` is a class in the `android.nfc.tech` package:

```
NdefFormatable ndefFormat = NdefFormatable.get(detectedTag);
```

In this part of the code, first the `ndefFormat` object is checked to see if it is null. If it is null it means that the tag does not support NDEF and you need to break the operation. If the tag supports NDEF, you continue:

```
if (ndefFormat != null) { … }
```

In order to enable I/O operations on the tag, the `connect()` method is applied on the instance of the tag. When the I/O operations are completed, the object should be closed with the `close()` method.

```
ndefFormat.connect();
…
ndefFormat.close();
```

Since the tag is not NDEF-formatted, you first need to format the tag as NDEF and then write the NDEF message. Android provides two operations with one method as shown here:

```
ndefFormat.format(message);
```

The `Ndef` and `NdefFormatable` classes also enable you to make a tag read-only. In the `Ndef` class, the `makeReadOnly()` method makes a tag read-only. In the `NdefFormatable` class, the `formatReadOnly(NdefMessage)` method makes a tag read-only by saving an NDEF message into it.

In order to write NDEF messages to tags, the described operations are sufficient. Now, you are able to write NDEF messages to tags, and the next section describes the how to read NDEF messages from tags.

TAG READING

As described earlier, when an application discovers a tag, it starts an activity that is defined in the `AndroidManifest.xml` file. Inside the activity you may perform different operations based on the discovered tag. This section describes how to read NDEF data from tags.

In the following code, the basic read operation from a tag is shown in which the discovered tag has NDEF payload:

```
if (NfcAdapter.ACTION_NDEF_DISCOVERED.equals(getIntent().getAction())){
    Tag detectedTag = getIntent().getParcelableExtra(NfcAdapter.EXTRA_TAG);

    // GET NDEF MESSAGES IN THE TAG

    // PROCESS NDEF MESSAGE

    // DO WHAT EVER YOU WANT WITH THE DATA
}
```

First, you check whether the intent was launched from an NFC interaction of `NDEF_DISCOVERED`:

```
if (NfcAdapter.ACTION_NDEF_DISCOVERED.equals(getIntent().getAction())){
```

Then, the instance of the discovered tag is obtained and saved to the `detectedTag` object:

```
Tag detectedTag = getIntent().getParcelableExtra(NfcAdapter.EXTRA_TAG);
```

Then, you need to get the NDEF messages from the tag, process the records inside the message, and do whatever you want with the application.

Getting an NDEF Message

In order to get the NDEF messages from the tag, you need to use the `NdefMessage` class. As shown in the following code, you first create an `NdefMessage` array to store the messages and call the `getNdefMessages` method to retrieve messages from the tag:

```
NdefMessage[] messages = getNdefMessages(getIntent());

NdefMessage[] getNdefMessages(Intent intent) {
    NdefMessage[] message = null;
    if (NfcAdapter.ACTION_NDEF_DISCOVERED.equals(intent.getAction())) {
        Parcelable[] rawMessages =
            intent.getParcelableArrayExtra(NfcAdapter.EXTRA_NDEF_MESSAGES);
        if (rawMessages != null) {
            message = new NdefMessage[rawMessages.length];
            for (int i = 0; i < rawMessages.length; i++) {
                message[i] = (NdefMessage) rawMessages[i];
            }
        } else {
            byte[] empty = new byte[] {};
            NdefRecord record = new NdefRecord ( NdefRecord.TNF_UNKNOWN,
```

```
                                          empty, empty, empty );
          NdefMessage msg = new NdefMessage( new NdefRecord[] { record } );
          message = new NdefMessage[] { msg };
        }
    } else {
        Log.d("", "Unknown intent.");
        finish();
    }
    return message;
}
```

Inside the `getNdefMessages` method, you retrieve the extended data from the intent by using the `intent.getParcelableArrayExtra` method. This method will get the data contained in the intent:

```
Parcelable[] rawMessages =
intent.getParcelableArrayExtra(NfcAdapter.EXTRA_NDEF_MESSAGES);
```

Then, a check is made of whether the `rawMessages` array is null for extended control. If it is not null, it means that the tag contains NDEF messages. Then, NDEF messages are saved into the message array, which is from the `NdefMessage` type:

```
if (rawMessages != null) {
    message = new NdefMessage[rawMessages.length];
    for (int i = 0; i < rawMessages.length; i++) {
        message[i] = (NdefMessage) rawMessages[i];
    }
}
```

If `rawMessages` equals null, it means that the tag is from an unknown type. In this situation, an empty message with the record of `TNF_UNKNOWN` is created and returned to the main program.

Processing an NDEF Message

In order to process the NDEF message, you need to know the contained record types. Since most of the record types need to be created in different ways, you should process them differently. In this section, you will process different record types, such as `TNF_WELL_KNOWN` with `RTD_URI`, `TNF_WELL_KNOWN` with `RTD_TEXT`, `TNF_ABSOLUTE_URI`, `TNF_MIME_MEDIA`, and `TNF_EXTERNAL_TYPE`.

Processing TNF_WELL_KNOWN with RTD_URI

As described in the "Preparing NDEF Data" section, `TNF_WELL_KNOWN` with `RTD_URI` includes the prefix in the first byte of the record, and the rest of the bytes in the record save the URL. So, for each record, you need to get the first byte and process it to find the corresponding prefix, and then append the rest of the bytes as the URL. Assume that the first byte of the record equals 1 and the rest of the bytes give `nfclab.com`. After processing the data in the tag, it is found that the corresponding prefix is 1, which means `"http://www."` Appending `nfclab.com` to the prefix gives the record of `"http://www. nfclab.com"`. The following code processes the records of an NDEF message:

```
for(int i=0;i<messages.length;i++){
    myText.append("Message "+(i+1)+":\n");
```

```
        for(int j=0;j<messages[0].getRecords().length;j++){
           NdefRecord record = messages[i].getRecords()[j];
           myText.append((j+1)+"th. Record Tnf: "+record.getTnf()+"\n");
           myText.append((j+1)+"th. Record type: " +
                             new String(record.getType())+"\n");
           myText.append((j+1)+"th. Record id: " +
                             new String(record.getId())+"\n");
           payload = new String(record.getPayload(), 1,
                          record.getPayload().length-1, Charset.forName("UTF-8"));
           myText.append((j+1)+"th. Record payload: "+payload+"\n");
           payloadHeader = record.getPayload()[0];
           myText.append((j+1)+"th. Record payload header: " +
              payloadHeader+"\n");
        }
     }
```

Remember that `messages` is an `NdefMessage` array that stores the messages by invoking the `getNdefMessages` method, which retrieves NDEF messages from the tag. As you can see from the record, first you start a loop to process all of the messages retrieved from the tag. Inside the loop, you start a second loop to process all of the records inside the message, because multiple records may exist in an NDEF message:

```
for(int i=0;i<messages.length;i++){
    ...
    for(int j=0;j<messages[0].getRecords().length;j++){
        ...
    }
}
```

For each record, you create an `NdefRecord` object to store the current record and get the `j`th record in the `i`th message with the `getRecords()` method:

```
NdefRecord record = messages[i].getRecords()[j];
```

Then, you print some data about the record to the screen, such as TNF, record type, and/or record ID. This step is not mandatory; however, it is good to know this information about the record. The important part is the step where you get the payload. The record's payload is retrieved with the `getPayload()` method. However, you need to split the payload into two parts: the first byte that stores the prefix and the rest of it that stores the URL. For this purpose, you save the record's bytes starting from index 1 to the end of the payload string variable and save its 0-indexed byte to the `payloadHeader` byte variable:

```
payload = new String( record.getPayload(), 1, record.getPayload().length-1,
                   Charset.forName("UTF-8"));
payloadHeader = record.getPayload()[0];
```

Now, you have the URL in a payload string (for example, `"nfclab.com"`), and the prefix in `payloadHeader` byte (for example, `0x01`). Furthermore, you need to process the `payloadHeader` in order to get the prefix and select the required operation. For example, if the `payloadHeader` value is

0x01 (http://www.) or 0x03 (http://), you may activate the browser to display the webpage. If the payloadHeader value is 0x06 (mailto:), you may run an e-mail program to send the e-mail. The following code describes the required operation for a prefix value of 0x01:

```
if(payloadHeader==0x01){
    Intent data = new Intent();
    data.setAction(Intent.ACTION_VIEW);
    data.setData(Uri.parse("http://"+payload));
    try {
        startActivity(data);
    } catch (ActivityNotFoundException e) {
    return;
    }
}
```

Processing TNF_WELL_KNOWN with RTD_TEXT

As described in the "Preparing NDEF Data" section, TNF_WELL_KNOWN with RTD_TEXT record payload includes a status byte, the language code of the text that can vary in size, and the actual text message. The required code to handle an RTD_TEXT record is as follows:

```
for(int i=0;i<messages.length;i++){
    myText.append("Message " + (i+1) + ":\n");
    for(int j=0;j<messages[0].getRecords().length;j++){
        NdefRecord record = messages[i].getRecords()[j];
        statusByte=record.getPayload()[0];
        int languageCodeLength = statusByte & 0x3F;
        myText.append("Language Code Length:" + languageCodeLength+"\n");
        String languageCode = new String( record.getPayload(), 1,
                            languageCodeLength, Charset.forName("UTF-8"));
        myText.append("Language Code:" + languageCode+"\n");
        int isUTF8 = statusByte-languageCodeLength;
        if(isUTF8 == 0x00){
            myText.append((j+1) + "th. Record is UTF-8\n");
            payload = new String( record.getPayload(), 1+languageCodeLength,
                record.getPayload().length-1-languageCodeLength,
                Charset.forName("UTF-8"));
        } else if (isUTF8==-0x80){
            myText.append((j+1) + "th. Record is UTF-16\n");
            payload = new String( record.getPayload(), 1+languageCodeLength,
                            record.getPayload().length-1-languageCodeLength,
                            Charset.forName("UTF-16"));
        }
        myText.append((j+1) + "th. Record Tnf: " + record.getTnf() + "\n");
        myText.append((j+1) + "th. Record type: " +
                            new String(record.getType()) + "\n");
        myText.append((j+1) + "th. Record id: " +
                            new String(record.getId()) + "\n");
        myText.append((j+1) + "th. Record payload: " + payload + "\n");
    }
}
```

In order to process an `RTD_TEXT` record payload, first you need to gather the record's first byte. When creating the `RTD_TEXT` record, remember that you have added the UTF bit and the language code length. In order to find out the length of the language code, you need to mask the record's first byte (status byte) with the value `0x3F`:

```
statusByte = record.getPayload()[0];
int languageCodeLength = statusByte & 0x3F;
```

Finding the length of the language code's length is important since the next bytes are language code. In order to find the language code, you need to get the record's bytes from 1 until the language code length. For example; if the language code's length is 3, then the bytes from 1 to 3 give the language code:

```
String languageCode = new String( record.getPayload(), 1,
    languageCodeLength, Charset.forName("UTF-8"));
```

Remember also that the text in `RTD_TEXT` records can be encoded either in UTF-8 or in UTF-16. In order to find the actual text's encoding, you need to process the status byte further. First, you subtract the language code length from the status byte:

```
int isUTF8 = statusByte-languageCodeLength;
```

If the resulting byte equals `0x00`, the text's encoding is UTF-8; if the resulting byte equals `0x80` then the encoding is UTF-16. After identifying the text encoding, the last thing to do is to process the actual text. The actual text starts from the bytes of (1+ the language code length) and lasts up to the end of the payload. For example, if the language code's length is 3 and the total size of the payload is 14, then the bytes from 4 to 13 give the actual text:

```
if(isUTF8 == 0x00){
    payload = new String( record.getPayload(), 1+languageCodeLength,
        record.getPayload().length-1-languageCodeLength, Charset.forName("UTF-8"));
} else if (isUTF8 == -0x80){
    payload = new String( record.getPayload(), 1+languageCodeLength,
        record.getPayload().length-1-languageCodeLength, Charset.forName("UTF-16"));
}
```

Processing TNF_ABSOLUTE_URI

In order to process a `TNF_ABSOLUTE_URI` record, all you need to do is to gather the record payload. The required code to handle `TNF_ABSOLUTE_URI` record is as follows:

```
for(int i=0;i<messages.length;i++){
    myText.append("Message "+(i+1)+":\n");
    for(int j=0;j<messages[0].getRecords().length;j++){
        NdefRecord record = messages[i].getRecords()[j];
        myText.append((j+1)+"th. Record Tnf: " + record.getTnf()+"\n");
        myText.append((j+1)+"th. Record type: " +
                        new String(record.getType())+"\n");
        myText.append((j+1)+"th. Record id: " + new String(record.getId())+"\n");
        payload = new String(record.getPayload());
        myText.append((j+1)+"th. Record payload: "+payload+"\n");
    }
}
```

Processing TNF_MIME_MEDIA

In order to process a `TNF_MIME_MEDIA` record, you need to gather the record payload similarly to the way it is done with the `TNF_ABSOLUTE_URI`. The required code to handle a `TNF_ABSOLUTE_URI` record is as follows:

```
for(int i=0;i<messages.length;i++){
    myText.append("Message "+(i+1)+"\n");
    for(int j=0;j<messages[0].getRecords().length;j++){
        NdefRecord record = messages[i].getRecords()[j];
        myText.append((j+1)+"th. Record Tnf: " + record.getTnf()+"\n");
        myText.append((j+1)+"th. Record type: " +
                            new String(record.getType()) + "\n");
        myText.append((j+1)+"th. Record id: " + new String(record.getId())+"\n");
        payload = new String(record.getPayload());
        myText.append((j+1)+"th. Record payload: "+payload+"\n");
    }
}
```

Processing TNF_EXTERNAL_TYPE

An NDEF record with a `TNF_EXTERNAL_TYPE` is handled in a similar way to the `TNF_MIME_MEDIA` and `TNF_ABSOLUTE_URI` records. The required code to handle a `TNF_EXTERNAL_TYPE` record is as follows:

```
for(int i=0;i<messages.length;i++){
    myText.append("Message "+(i+1)+":\n");
    for(int j=0;j<messages[0].getRecords().length;j++){
        NdefRecord record = messages[i].getRecords()[j];
        myText.append((j+1)+"th. Record Tnf: " + record.getTnf()+"\n");
        myText.append((j+1)+"th. Record type: " +
                            new String(record.getType())+"\n");
        myText.append((j+1)+"th. Record id: " + new String(record.getId())+"\n");
        payload = new String(record.getPayload());
        myText.append((j+1)+"th. Record payload: "+payload+"\n");
    }
}
```

ANDROID APPLICATION RECORD

Android Application Record (AAR) is introduced in API level 14 and is a very powerful property of Android NFC. The main objective of AAR is to start the application when an NFC tag is scanned. Despite the fact that the manifest file gives this property to NFC, AAR strengthens the property because more than one application may have been registered for, and may filter, the same intents on a mobile phone. Additionally, AAR helps users to download and install the application automatically after scanning NFC tags.

An AAR is the package name of an application. You need to insert the AAR record to an NDEF message, similarly to the way it is done with an NDEF record.

How It Works

When an NFC tag is scanned, Android searches the entire NDEF message for AAR. If it finds an AAR in any of the NDEF records, it starts the corresponding application. If the application is not installed on the device, Google Play is launched automatically to download the corresponding application.

If the activity that filters the intent does not match the AAR, or if multiple or no activities handle the intent, it does not matter and the application specified by the AAR is started.

Intent Filters vs. AAR

Intent filters and AAR have some differences. First of all, AARs are supported at application level; however, intent filters are supported at activity level. The reason for this is that AAR uses the package name and cannot be lowered to the activity level. Thus, if you want to handle the intent at activity level for any reason, you need to use intent filters.

The second difference is that there can be only one application that matches with an AAR; however, there can be many applications registered to handle intents. Thus, using AAR is more powerful for running applications automatically.

Finally, AAR can also be used to download applications automatically from Google Play if no application with the same package name is installed. However, if there is no application registered to handle the discovered intent, nothing is performed.

Important Notes on AAR

First of all, note that you can use AAR and intent filters together. If you want your application to handle only the tags you deployed, you may use AAR without using intent filters. However, in this situation, note that AAR will not work with the devices that have an API level lower than 14. If you wish to deploy applications to the devices that have lower API levels, you may use intent filters together with AAR.

If you want your application to handle the tags that you deployed as well as others (such as tags with a URI or specific MIME type), you must use intent filters. However, you can also use AAR on your tags so that your application will run automatically when your tags that contain AAR are scanned.

Using AAR

In order to create an AAR record inside an NDEF message, use the `NdefRecord` `.createApplicationRecord` method and give your package name as a parameter in this method. You may perform this operation when creating the NDEF message with NDEF records. The following example demonstrates the usage:

```
NdefMessage newMessage= new NdefMessage(new NdefRecord[] {
    record1, record2, ... ,
    NdefRecord.createApplicationRecord("com.nfclab.smartposter")
});
```

When reading the tag, you do not need any additional codes, since the operation is performed in application level, not in activity level.

FOREGROUND DISPATCH SYSTEM

As described at the beginning of the chapter, the foreground dispatch system handles the tags when the application is in active state. It simply gives the priority to handle the tag to the application that is running.

In order to implement the foreground dispatch system, you first need to create a `PendingIntent` so that Android can get the details of the tag:

```
PendingIntent pendingIntent = PendingIntent.getActivity(
        this, 0, new Intent(this,
                getClass()).addFlags(Intent.FLAG_ACTIVITY_SINGLE_TOP), 0);
```

Then you need to declare the intent filters to handle the tags. You can think of these intent filters as being the same as the intent filters that are defined in manifest files in the tag intent dispatch system. If the registered filters in the foreground dispatch system match with the tag, then the active application handles the intent. If not, then the intent dispatch system looks for the available activities that can handle the tag and runs the corresponding activity that already registered the intent that can handle the tag.

In order to register for `NDEF_DISCOVERED`, use the following code:

```
IntentFilter ndef = new IntentFilter(NfcAdapter.ACTION_NDEF_DISCOVERED);
try {
   ndef.addDataType("*/*");
}
catch (MalformedMimeTypeException e) {
   throw new RuntimeException("fail", e);
}
intentFiltersArray = new IntentFilter[] {ndef, };
```

The code handles all MIME types, since the `addDataType` added a new intent data type of `*/*` to match against the discovered NDEF data. For example, in order to handle `RTD_URI`, you need to use the `addDataScheme` method like this:

```
ndef.addDataScheme("http");
```

You can also specify a set of tag technologies to handle `ACTION_TECH_DISCOVERED` intents. You need to use the format of the `Object.class.getName()` method to obtain the tag technologies that you want to support. The following code adds the `NfcF` tag technology in order to filter it:

```
techListsArray = new String[][] { new String[] { NfcV.class.getName() } };
```

The following code adds `MifareClassic` and `NfcA` tag types to filter:

```
techListsArray = new String[][] { new String[] {
                MifareClassic.class.getName(), NfcA.class.getName() } };
```

In order to add multiple tech-lists, you may use the format in the following example. It adds a tech-list with the `NfcV` tag type and another list with the `MifareClassic` and `NfcA` tag types:

```
techListsArray = new String[][] {
        new String[] { NfcV.class.getName() },
        new String[] { MifareClassic.class.getName(),
                NfcA.class.getName() } };
```

Finally, you need to override the following activity lifecycle callbacks and add logic to enable and disable the foreground dispatch on onPause() and onResume():

```
public void onPause() {
   super.onPause();
   NFCAdapter.disableForegroundDispatch(this);
}

public void onResume() {
   super.onResume();
   NFCAdapter.enableForegroundDispatch(this, pendingIntent,
                                intentFiltersArray, techListsArray);
}

public void onNewIntent(Intent intent) {
   Tag tag = intent.getParcelableExtra(NfcAdapter.EXTRA_TAG);
}
```

You need to implement the required code to process the data from the scanned NFC tag in the onNewIntent method. The following code is the implementation of reading TNF_WELL_KNOWN with RTD_URI inside the onNewIntent method:

```
Tag tag = intent.getParcelableExtra(NfcAdapter.EXTRA_TAG);
NdefMessage[] messages = getNdefMessages(intent);
for(int i=0;i<messages.length;i++){
   for(int j=0;j<messages[0].getRecords().length;j++){
      NdefRecord record = messages[i].getRecords()[j];
      payload = new String(record.getPayload(), 1, record.getPayload().length-1,
         Charset.forName("UTF-8"));
      payloadHeader = record.getPayload()[0];
   }
}

if(payloadHeader == 0x01){
   Intent data = new Intent();
   data.setAction(Intent.ACTION_VIEW);
   data.setData(Uri.parse("http://"+payload));
   try {
      startActivity(data);
   } catch (ActivityNotFoundException e) {
      return;
   }
}
```

WORKING WITH SUPPORTED TAG TECHNOLOGIES

In NFC, you generally need to use the NDEF format to write and read to and from tags. However, when a tag is scanned, the tag may not be NDEF-formatted or your application may not be registered to handle the tag's NDEF format. In these situations you should access the tag in raw bytes and perform the required operations. The tag types to which Android gives access are provided in the android.nfc.tech package and are also listed in Table 5-4 and Table 5-5. You can retrieve the type of the tag using the getTechList() method and then perform the communication to the tag with the classes provided in the android.nfc.tech package.

TABLE 5-4: Supported Tag Technologies

TAG TYPE	CLASS	DESCRIPTION
TagTechnology	`android.nfc.tech.` `TagTechnology`	The interface that all tag technology classes must implement
NfcA	`android.nfc.tech.NfcA`	Implements operations on NFC-A (ISO 14443-3A) tags
NfcB	`android.nfc.tech.NfcB`	Implements operations on NFC-B (ISO 14443-3B) tags
NfcF	`android.nfc.tech.NfcF`	Implements operations on NFC-F (JIS 6319-4) tags
NfcV	`android.nfc.tech.NfcV`	Implements operations on NFC-V (ISO 15693) tags
IsoDep	`android.nfc.tech.IsoDep`	Implements operations on ISO-DEP (ISO 14443-4) tags
Ndef	`android.nfc.tech.Ndef`	Implements NDEF operations on tags
NdefFormatable	`android.nfc.tech.` `NdefFormatable`	Implements format operations for tags that can be NDEF formattable
NfcBarcode	`android.nfc.tech.NfcBarcode`	Provides access to tags containing just a barcode

TABLE 5-5: RTD_TEXT Payload

TAG TYPE	CLASS	DESCRIPTION
MifareClassic	`android.nfc.tech.MifareClassic`	Implements operations on MIFARE Classic tags
MifareUltralight	`android.nfc.tech.` `MifareUltralight`	Implements operations on MIFARE Ultralight tags

Getting Available Tag Technologies

In order to connect to connect to a tag, you need to know the type of the tag. If you wish to get the available tag types that a tag owns, use the `getTechList()` method as shown in the following code:

```
if (NfcAdapter.ACTION_TECH_DISCOVERED.equals(getIntent().getAction())) {
    Tag tag = getIntent().getParcelableExtra(NfcAdapter.EXTRA_TAG);
    myText.setText("Technologies available in this tag=" +
        Arrays.toString(tag.getTechList()) );
}
```

After getting the supported tag technologies of a tag, you may connect to tag one of these technologies. For example, if a scanned tag supports NfcV and NdefFormatable, then you can get the tag with either of the classes (`android.nfc.tech.NfcV` or `android.nfc.tech.NdefFormatable`) and perform the required I/O operations.

NfcV Example

In this example, you will implement I/O operations on ISO 15693 type (NfcV) tags. First of all, your application should include the required codes in the manifest file to filter for ACTION_TECH_DISCOVERED intents. The following code is the required code in order to filter ACTION_TECH_DISCOVERED intents with a predefined tag technology list of `nfc_tech_list.xml`:

```
<activity>
...
    <intent-filter>
        <action android:name="android.nfc.action.TECH_DISCOVERED"/>
    </intent-filter>
    <meta-data android:name="android.nfc.action.TECH_DISCOVERED"
        android:resource="@xml/nfc_tech_list"/>
</activity>
```

The XML file includes the NfcV tag type that handles the I/O operations on it:

```
<?xml version="1.0" encoding="utf-8"?>
<resources xmlns:xliff="urn:oasis:names:tc:xliff:document:1.2">
    <tech-list>
        <tech>android.nfc.tech.NfcV</tech>
    </tech-list>
</resources>
```

After including the required code in the manifest file and creating the tag technology list XML file, you need to implement the required code to perform I/O operations. The following code implements a read operation on byte block 10 of the NfcV tags:

```
if (NfcAdapter.ACTION_TECH_DISCOVERED.equals(getIntent().getAction()))
{
    Tag detectedTag = getIntent().getParcelableExtra(NfcAdapter.EXTRA_TAG);
    NfcV nfcv = NfcV.get(detectedTag);
    try {
        nfcv.connect();
        if(nfcv.isConnected()){
            myText.append("Connected to the tag");
            myText.append("\nTag DSF: "+Byte.toString(nfcv.getDsfId()));
            byte[] buffer;
            buffer=nfcv.transceive(new byte[] {0x00, 0x20, (byte) 10});
            myText.append("\nByte block 10:"+buffer);
            myText.append("\nByte block 10 as string:"+new String(buffer));
            nfcv.close();
        } else
            myText.append("Not connected to the tag");
    } catch (IOException e) {
        myText.append("Error");
    }
}
```

In the example, you first need to check if the tag is discovered with the `ACTION_TECH_DISCOVERED` intent:

```
if (NfcAdapter.ACTION_TECH_DISCOVERED.equals(getIntent().getAction()))
{
    ...
}
```

Then you need to obtain the tag object from the intent:

```
Tag detectedTag = getIntent().getParcelableExtra(NfcAdapter.EXTRA_TAG);
```

You need to get an instance of the `TagTechnology` (in the example it is NfcV) by calling the corresponding `get()` method in the tag technology's package:

```
NfcV nfcv = NfcV.get(detectedTag);
```

The example implements NfcV tags only; however, your application may handle more than one tag type. In this situation, you can get the supported technologies in the tag by using the `getTechList()` method before obtaining the instance of the `TagTechnology`. You should first learn the supported technology and then obtain the instance of the `TagTechnology` based on the supported types.

After this part of the code, each tag technology has different methods and each tag type needs different implementations. You may implement any method defined in the tag technology's class.

For NfcV tags, you may implement any method in `android.nfc.tech class.NfcV`. In order to perform operations on NfcV tags, you need to first connect to it:

```
nfcv.connect();
```

In order to get the tag's DSF ID in bytes from tag discovery, you need to use the following:

```
nfcv.getDsfId()
```

In order to send commands to the NfcV tags, you need to use the `transceive` method. When writing data to and reading data from NfcV tags, the operation is performed on the byte blocks that consist of 4 bytes. Reading data from NfcV tags is `0x20` and writing is `0x21`. In order to read the 10th byte block on the tag, the following code is required:

```
buffer=nfcv.transceive(new byte[] {0x00, 0x20, (byte) 10});
```

You may either print the data in bytes or in a string:

```
myText.append("\nByte block 10:"+buffer);
myText.append("\nByte block 10 as string:"+new String(buffer));
```

In order to write data to NfcV tags, you need to use `0x21`. The following code writes the bytes of `0x00, 0x00, 0x72, 0x75` to the data 10th byte block:

```
Buffer = nfcv.transceive(new byte[] {0x00, 0x21, (byte) 10,0x00,
                                     0x00, 0x72, 0x75});
```

The classes in the `android.nfc.tech` package provide access to the tags and I/O operations in raw bytes. Each tag technology has different implementations for I/O operations. You need to be well aware of the implemented tag technology, the data structure of the tag, and so on, to use the corresponding classes and to program applications for specific tag types.

SUMMARY

There are two packages for NFC application development in an Android platform. The first package is `android.nfc`, which provides mobile phones to read/write NDEF messages from/to supported tags. This package also enables data exchange with other NFC-enabled mobile phones. There are six classes in this package: `Tag`, `NfcAdapter`, `NfcManager`, `NfcEvent`, `NdefMessage`, and `NdefRecord`.

The other package is the `android.nfc.tech` package, which includes necessary classes to provide access to different tag technologies such as MIFARE Classic, NfcA, NfcV, and so on. This package also provides I/O operations on these tags in raw bytes, and includes the classes required by different tag types in order to perform I/O operations. Some of the classes are `IsoDep`, `MifareClassic`, `MifareUltralight`, and `NfcV`. The package also includes a `TagTechnology` interface to obtain the tag and connect to it.

NFC APIs are introduced from API level 9, including the `android.nfc` package. However, most of the classes and methods in this package are introduced in API level 10 and more are added in API level 14. So you should use at least API level 10 to provide NFC capabilities. Refer to Appendix B for a detailed list of packages, classes, and constants.

The tag intent dispatch system is used to launch applications when a tag or a specific NDEF record is identified in a tag. On the other hand, the foreground dispatch system is designed to handle tags when the application is running. The difference in the coding is that the tag intent dispatch system registers the tag types and NDEF data that the application can handle in the application's manifest file using intent filters, whereas the foreground dispatch system registers inside the activity.

There are three intents: `ACTION_NDEF_DISCOVERED`, `ACTION_TECH_DISCOVERED`, and `ACTION_TAG_DISCOVERED`. `ACTION_NDEF_DISCOVERED` is used when the NDEF payload can be mapped to a MIME type or URI. `ACTION_TECH_DISCOVERED` is used when the NDEF payload cannot be mapped to a MIME type or URI. `ACTION_TAG_DISCOVERED` is used when a tag does not contain an NDEF payload but is one of the known tag technologies.

Supported TNFs in Android are `TNF_ABSOLUTE_URI`, `TNF_EMPTY`, `TNF_EXTERNAL_TYPE`, `TNF_MIME_MEDIA`, `TNF_UNCHANGED`, `TNF_UNKNOWN`, and `TNF_WELL_KNOWN`.

In order to enable an application to use the NFC adapter, you should permit the application to use the NFC hardware in the Android manifest file.

The main objective of Android Application Record (AAR) is to start the application when an NFC tag is scanned. Despite the fact that the manifest file gives this property using intent filters, more than one application may have been registered to filter the same intents on a mobile phone. However, there can be only one application that matches with an AAR, so AAR guarantees an application will run when a specified tag is scanned. Additionally, AAR helps users to download and install the application automatically after scanning NFC tags, if it is not installed in the mobile.

Reader/Writer Mode Applications

WHAT'S IN THIS CHAPTER?

➤ Use cases for reader/writer mode applications

➤ Programming an NFC Smart Poster use case

➤ Programming an NFC shopping use case

➤ Programming an NFC student transportation tracking use case

WROX.COM CODE DOWNLOADS FOR THIS CHAPTER

The wrox.com code downloads for this chapter are found at `www.wrox.com/remtitle .cgi?isbn=1118380096` on the Download Code tab. The code is in the Chapter 6 download and individually named according to the names throughout the chapter.

In this chapter, three different NFC use cases are programmed using the Android APIs. All three cases use reader/writer mode. For all cases, main applications and tag writer applications are programmed separately.

The first use case consists of a Smart Poster scenario, and includes the following subcases:

➤ Displaying a webpage

➤ Calling a phone number

➤ Sending an SMS

➤ Displaying a geolocation

➤ Sending an e-mail

The second use case consists of an NFC-based shopping scenario, which enables mobile shopping by touching preformatted NFC tags. When the mobile phone is touched to an NFC tag, the corresponding product is added to the basket.

The third use case enables tracking student transportation activities between school and home. When a student gets on or off a school bus, they touch their NFC tag to the mobile phone in the school bus and the mobile updates that student's status, including the times when the student gets on and off the bus.

NFC SMART POSTER USE CASE

Smart Poster, which makes a poster smart by integrating NFC tags to it, is one of the most popular use cases of NFC reader/writer mode. One or more NFC tags may be placed onto the poster and different data types may be saved into those NFC tags to trigger different services on the mobile phone. For example, both an NFC tag containing a webpage URL and an NFC tag containing an e-mail address can be placed onto a poster to trigger corresponding activities on the mobile.

In the design phase, it is very important to decide which applications can read the NFC tag. One option is to allow all mobile phones to read the tag. In this case, when an NFC mobile scans the Smart Poster, Android OS handles the data itself. Another option is to allow only one specific application to handle the tag. The TNF and RTD of the NDEF record to be written to the tag need to be selected based on the design criteria. If you use a TNF_WELL_KNOWN record, then all mobile devices are able to handle the record. However, if you select a TNF_EXTERNAL_TYPE record, only a specific application can use the data on the tag. In this Smart Poster application, the tags for three subcases (webpage link, phone number, and e-mail) are programmed with TNF_WELL_KNOWN records, whereas TNF_EXTERNAL_TYPE is used for SMS and geolocation scenarios with the external types of nfclab.com:smsService and nfclab.com:geoService. Two different applications are developed for the Smart Poster use case. The first program is the writer application that enables coding of the tags for five different data types used in the use cases. The second one is the main program that reads tags and performs action on the received data.

Smart Poster Tag Writer Application

This application is used to write the content of the NFC tags that will be placed on Smart Poster. The main screen of the application is shown in Figure 6-1, which includes the necessary buttons for the five different data types that will be used in this application.

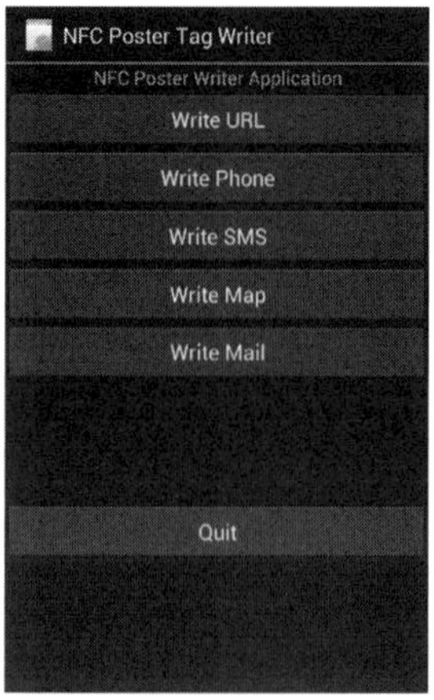

The main activity of the application is NFCPosterActivity. GridLayout is used as the activity's layout. The layout includes five buttons as menu items, and another button to quit the application. onClick listeners are implemented for each button using the setOnClickListener() method. In each method except the "Quit" button, a new activity is invoked using intents. The names of those activities are as follows:

FIGURE 6-1

➤   `WriteUrlActivity`

➤   `WritePhoneActivity`

➤   `WriteSmsActivity`

➤   `WriteMapActivity`

➤   `WriteMailActivity`

Writing a URL to an NFC Tag

When the "Write URL" button is clicked, `WriteUrlActivity` is called and the screen shown in Figure 6-2 is displayed. This screen includes an `EditText` to input the webpage link, and a "Save to Tag" button to save the text to the NFC tag. When the URL is written and the "Save to Tag" button is clicked, a message is displayed indicating that the device is ready to save the data to the tag (see Figure 6-3). As the mobile phone scans the tag, the data is saved to the tag and then the screen shown in Figure 6-4 is displayed. If the write operation is unsuccessful for any reason, an error message is displayed. Most of the unsuccessful operations happen when the tag's size is not large enough to save the data. See Figure 6-5 for the activity flow diagram of the `WriteURLActivity`.

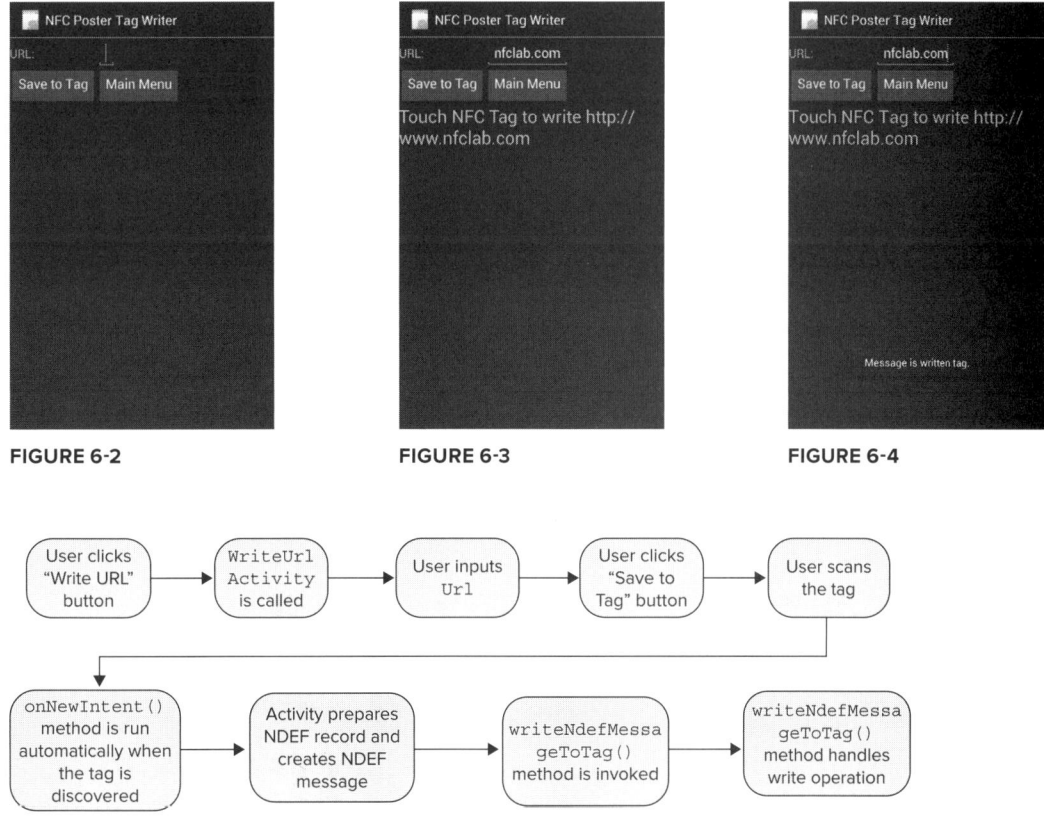

FIGURE 6-2

FIGURE 6-3

FIGURE 6-4

FIGURE 6-5

In the activity, the user inputs the URL to the `EditText` and then clicks the "Save to Tag" button. This button has a listener when it is clicked, which is implemented via the `setOnClickListener()` method. When this button is clicked, the URL in `EditText` is saved to the `urlAddress` string and a message is displayed telling the user to touch the tag. The code for this activity is as follows:

```
writeUrlButton.setOnClickListener( new View.OnClickListener() {
    public void onClick(View view) {
        urlAddress = urlEditText.getText().toString();
        TextView messageText = (TextView)findViewById(R.id.messageText);
        messageText.setText("Touch NFC Tag to write http://www."+urlAddress);
    }
});
```

The rest of the code is related to NFC technology. The foreground dispatch system is implemented in this and other activities of this application. First you need to create an `NfcAdapter` object to get the phone's NFC adapter:

```
NfcAdapter mNfcAdapter = NfcAdapter.getDefaultAdapter(this);
```

As described in the previous chapter, in order to implement the foreground dispatch system, first you need to create a `PendingIntent` so that the activity can get the details of the tag:

```
PendingIntent mPendingIntent = PendingIntent.getActivity(this, 0,
  new Intent(this, getClass()).addFlags(Intent.FLAG_ACTIVITY_SINGLE_TOP), 0);
```

Then you need to declare the intent filters to handle the tags and the tag types that your application can handle:

```
IntentFilter ndef = new IntentFilter(NfcAdapter.ACTION_NDEF_DISCOVERED);
mFilters = new IntentFilter[] {ndef, };
mTechLists = new String[][] { new String[] { Ndef.class.getName() },
new String[] { NdefFormatable.class.getName() }};
```

Then to enable and disable the foreground dispatch on `onPause()` and `onResume()`, two methods are implemented. The `disableForegroundDispatch()` method in `onPause()` disables the foreground dispatch system temporarily, and the `enableForegroundDispatch()` method enables the foreground dispatch system again with the given parameters of pending intent, intent filters, and tag technology types:

```
@Override
public void onPause() {
    super.onPause();
    mNfcAdapter.disableForegroundDispatch(this);
}

@Override
public void onResume() {
    super.onResume();
    if (mNfcAdapter != null) {
        mNfcAdapter.enableForegroundDispatch(this, mPendingIntent,
                                    mFilters, mTechLists);
    }
}
```

In order to write the data to the tag when it is discovered, you need to implement the `onNewIntent()` method, which is shown in the following code snippet:

```
@Override
public void onNewIntent(Intent intent) {
    Tag tag = intent.getParcelableExtra(NfcAdapter.EXTRA_TAG);
    byte[] uriField = urlAddress.getBytes(Charset.forName("US-ASCII"));
    byte[] payload = new byte[uriField.length + 1];
    payload[0] = 0x01;
    System.arraycopy(uriField, 0, payload, 1, uriField.length);
    NdefRecord URIRecord = new NdefRecord(NdefRecord.TNF_WELL_KNOWN,
                                    NdefRecord.RTD_URI, new byte[0], payload);
    NdefMessage newMessage= new NdefMessage(new NdefRecord[] { URIRecord });
    writeNdefMessageToTag(newMessage, tag);
}
```

In this activity, you format the data that will be written to the tag as `TNF_WELL_KNOWN` with `RTD_URI`. At first, the properties of the tag are received from incoming intent and saved to a tag object. Then, payload data is prepared by encoding the inputted data using the US-ASCII charset. Afterwards, the header of the NDEF record is prepared according to the `RTD_URI` specification. Since `http://www.` will be used as the prefix for the data, the first byte of the header is stored as `0x01` (please see Appendix A for prefixes). Then the payload and the header are merged into a payload byte array and an NDEF record is created from this byte array with the `RTD_URI` record type. Since you need only one NDEF record for this tag, a new NDEF message is composed using this record by calling the `writeNdefMessageToTag()` method. Please see Chapter 5, "NFC Programming: Reader/Writer Mode," for a detailed description of the `writeNdefMessageToTag()` method.

Writing a Phone Number to an NFC Tag

When the "Write Phone" button is clicked, the screen shown in Figure 6-6 is displayed. The logical model is the same as the URL option. When the "Save to Tag" button is clicked, a message is displayed to direct the user to touch the device to the tag (see Figure 6-7). Finally, when the mobile device scans the tag and the data is successfully written to the tag, a confirmation message is displayed on the screen (see Figure 6-8). See Figure 6-9 for the activity flow diagram of the `WritePhoneActivity`.

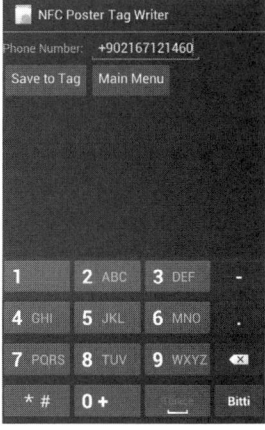

FIGURE 6-6

FIGURE 6-7

FIGURE 6-8

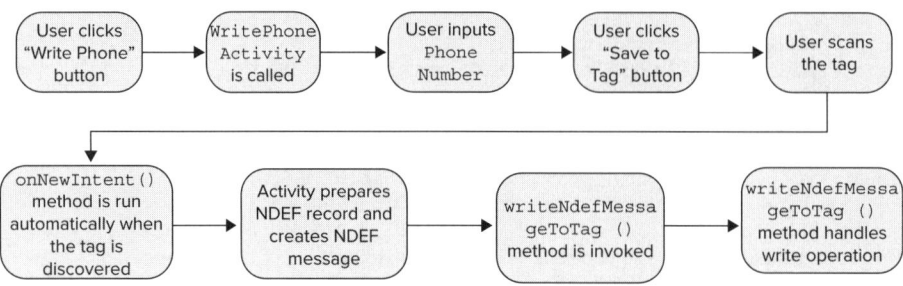

FIGURE 6-9

The majority of the code to write a phone number to the NFC tag is the same as the code for writing a URL to NFC tags. The only difference is in the onNewIntent() method, since the URI prefix of the NDEF message needs to be changed. Please remember that in order to write URLs to tags, you used the prefix 0x01 (http://www); however, in the case of phone numbers, you need to use the tel: prefix so that the first byte needs to be created as 0x05 (please see Appendix A for prefixes). The complete onNewIntent() method is shown here:

```
@Override
public void onNewIntent(Intent intent) {
    Log.i("Foreground dispatch", "Discovered tag with intent: " + intent);
    Tag tag = intent.getParcelableExtra(NfcAdapter.EXTRA_TAG);
    byte[] uriField = urlAddress.getBytes(Charset.forName("US-ASCII"));
    byte[] payload = new byte[uriField.length + 1];
    payload[0] = 0x05;
    System.arraycopy(uriField, 0, payload, 1, uriField.length);
    NdefRecord URIRecord = new NdefRecord(NdefRecord.TNF_WELL_KNOWN,
                                          NdefRecord.RTD_URI, new byte[0], payload);
    NdefMessage newMessage = new NdefMessage(new NdefRecord[] { URIRecord });
    writeNdefMessageToTag(newMessage, tag);
}
```

Writing an SMS Message to an NFC Tag

When the "Write SMS" button is clicked, the screen shown in Figure 6-10 is displayed. The same logical model applies here, so that when the "Save to Tag" button is clicked and the mobile device scans the tag, the data is written to the tag. Finally, a confirmation or error message is displayed on the screen (see Figure 6-11). See Figure 6-12 for the activity flow diagram of the WriteSmsActivity.

FIGURE 6-10

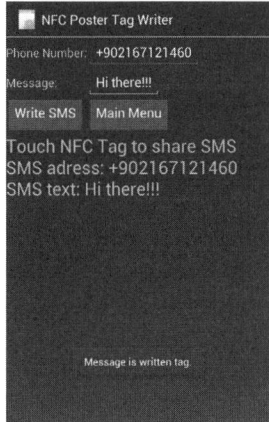

FIGURE 6-11

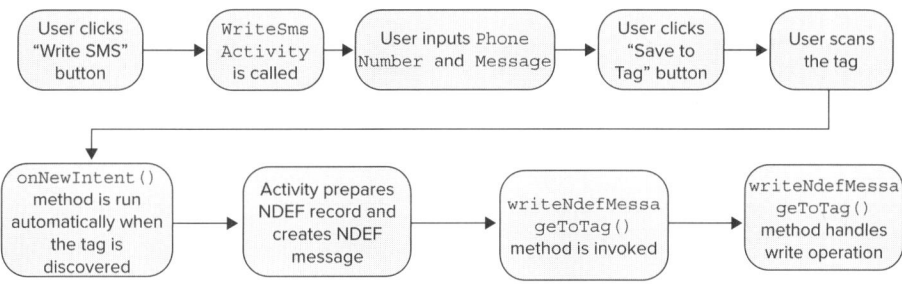

FIGURE 6-12

In order to write an SMS message to a tag, two different texts are needed: the phone number and the SMS body. So, two `EditText` objects are created and used in this case. The `smsNumber` and `smsBody` strings get the phone number and SMS body from the user. The complete code of the `setOnClickListener()` method, including these strings, is as follows:

```
writeSmsButton.setOnClickListener(new android.view.View.OnClickListener() {
    public void onClick(View view) {
        String smsNumber = smsNumberEditText.getText().toString();
        String smsBody = smsBodyEditText.getText().toString();
        urlAddress = "sms:"+smsNumber+"?body="+smsBody;
        TextView messageText = (TextView)findViewById(R.id.messageText);
        messageText.setText("Touch NFC Tag to share SMS\n" +
                    "SMS adress: "+smsNumber+"\nSMS text: "+smsBody);
    }
});
```

In order to create a URI for an SMS that includes a phone number and text, you need to use the following format:

```
sms:<SMS number>?body=<SMS body>
```

For example, a URI to constitute an SMS for the following information,

> Phone number: +902167121460
> SMS Body: Hello there

needs to be set as:

```
sms:+ 902167121460?body=Hello there
```

Thus, the `urlAddress` string given in the `setOnClickListener` method is set as follows:

```
urlAddress = "sms:"+smsNumber+"?body="+smsBody;
```

In this example, the `TNF_EXTERNAL_TYPE` record type needs to be used instead of `TNF_WELL_KNOWN`, because the NFC Forum has not defined any URI identifier code for `sms:` prefix yet (see Appendix A). Everything remains the same except the `onNewIntent()` method, as shown in the following code. In this method, an NDEF record is created with a `TNF_EXTERNAL_TYPE` named `nfclab.com:smsService`. Then an NDEF message is created from this NDEF record and the `writeNdefMessageToTag()` method is called, which enables writing the message to NFC tags.

```
@Override
public void onNewIntent(Intent intent) {
    Log.i("Foreground dispatch", "Discovered tag with intent: " + intent);
    Tag tag = intent.getParcelableExtra(NfcAdapter.EXTRA_TAG);
    String externalType = "nfclab.com:smsService";
    NdefRecord extRecord = new NdefRecord(NdefRecord.TNF_EXTERNAL_TYPE,
                                externalType.getBytes(), new byte[0],
                                urlAddress.getBytes());
    NdefMessage newMessage = new NdefMessage(new NdefRecord[] { extRecord});
    writeNdefMessageToTag(newMessage, tag);
}
```

Writing a Geolocation to an NFC Tag

When the "Write Map" button is clicked, the screen shown in Figure 6-13 is displayed. When the geolocation data are successfully saved to the tag, the screen shown in Figure 6-14 is displayed. See Figure 6-15 for the activity flow diagram of the `WriteMapActivity`.

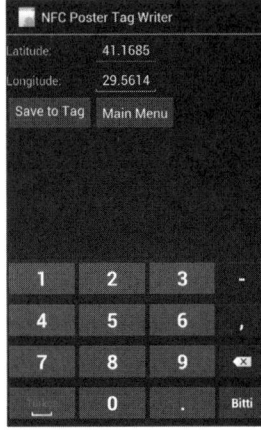

FIGURE 6-13

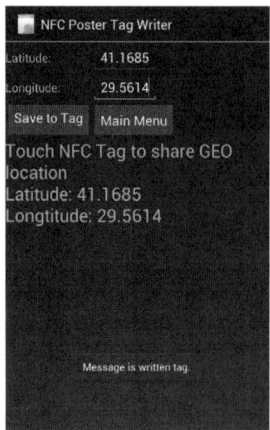

FIGURE 6-14

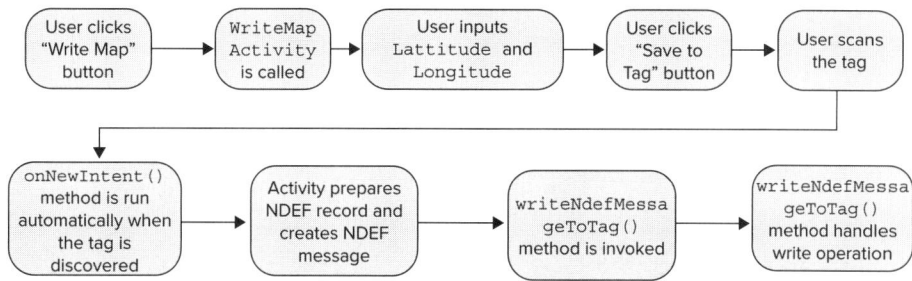

FIGURE 6-15

In order to write a map location to a tag, two different texts are needed. The first is `latitude` and the second is `longitude`. `latitude` and `longitude` strings store the `latitude` and `longitude` data inputted by the user via `EditText` objects. The following code shows the complete `setOnClickListener()` method, which includes the strings:

```
writeMapButton.setOnClickListener(new android.view.View.OnClickListener() {
    public void onClick(View view) {
        String latitude = latitudeEditText.getText().toString();
        String longitude = longitudeEditText.getText().toString();
        urlAddress = "geo:" + latitude + "," + longitude;
        TextView messageText = (TextView)findViewById(R.id.messageText);
        messageText.setText("Touch NFC Tag to share GEO location\n"+
                    "Latitude: " + latitude + "\nLongitude: " + longitude);
    }
});
```

In order to create a URI for a geolocation that includes the phone number and text, you need to use the following format:

```
"geo:"<lat>","<lon>"
```

For example, a URI to constitute a `geo` for the following information,

Latitude: 41.168898
Longitude: 29.564281

needs to be set as:

```
geo: 41.168898, 29.564281
```

Thus, the `urlAddress` string given in the `setOnClickListener()` method is set as follows:

```
urlAddress = "geo:" + latitude + "," + longtitude;
```

The `TNF_EXTERNAL_TYPE` record type is used in this example. Similar to the preceding SMS example, this record type is used instead of `TNF_WELL_KNOWN`, because NFC Forum has not defined the URI identifier code for the `geo:` prefix (see Appendix A). In the `onNewIntent()` method, an NDEF record is created with `TNF_EXTERNAL_TYPE` and named as `nfclab.com:geoService`. Then an NDEF message is created from this NDEF record, as shown in the following code:

```
@Override
public void onNewIntent(Intent intent) {
    Log.i("Foreground dispatch", "Discovered tag with intent: " + intent);
```

```
        Tag tag = intent.getParcelableExtra(NfcAdapter.EXTRA_TAG);
        String externalType = "nfclab.com:geoService";
        NdefRecord extRecord = new NdefRecord(NdefRecord.TNF_EXTERNAL_TYPE,
                                        externalType.getBytes(), new byte[0],
                                        urlAddress.getBytes());
        NdefMessage newMessage = new NdefMessage(new NdefRecord[] { extRecord});
        writeNdefMessageToTag(newMessage, tag);
    }
```

Writing an E-mail to an NFC Tag

When the "Write Mail" button is clicked, the screen shown in Figure 6-16 is displayed. When the user inputs the e-mail address, subject, and body, and clicks the "Save to Tag" button, the mobile waits for the tag. When the mobile device scans the tag, the data is written to the tag (see Figure 6-17). See Figure 6-18 for the activity flow diagram of the WriteMailActivity.

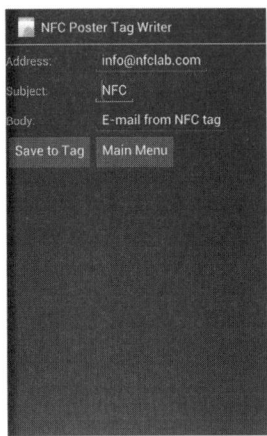

FIGURE 6-16

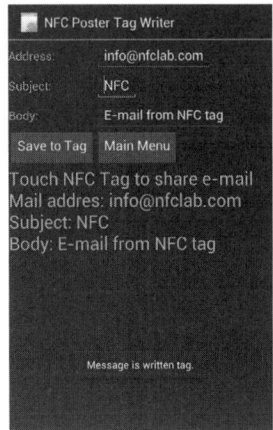

FIGURE 6-17

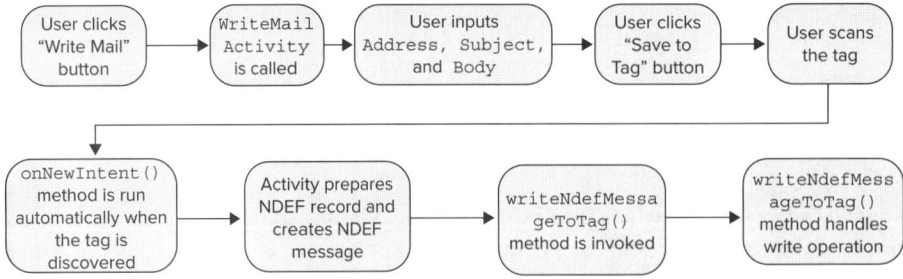

FIGURE 6-18

Since you need to get three different texts from the user via EditText objects, you need to create three different strings to handle the entered texts. mailAddress, mailSubject, and mailBody strings get the address field, subject field, and body field of the e-mail respectively.

In order to create a URI for an e-mail that includes its subject and body, you need to use the following format:

```
mailto:<address>[?<header1>=<value1>[&<header2>=<value2>]]
```

For example, if you need to create a URI to constitute e-mail for the following information:

E-mail address (mailto): info@nfclab.com
Subject: NFC
Body: E-mail from NFC Tag

You need to set the e-mail URI as:

```
mailto: info@nfclab.com?subject=NFC&body=E-mail from NFC Tag
```

Thus, the urlAddress string given in the setOnClickListener() method is set as follows:

```
urlAddress = mailAddress + "?subject=" + mailSubject + "&body=" + mailBody;
```

Remember that the urlAddress string will be used in the NDEF record's payload and the mailto: prefix will be set as the NDEF record's first byte as 0x06 (see Appendix A). So, you should not include mailto: in the urlAddress string. The following code shows the setOnClickListener() method:

```
writeMailButton.setOnClickListener(new android.view.View.OnClickListener() {
    public void onClick(View view) {
        String mailAddress = addressEditText.getText().toString();
        String mailSubject = subjectEditText.getText().toString();
        String mailBody = mailBodyEditText.getText().toString();
        urlAddress = mailAddress+"?subject=" +mailSubject+"&body="+mailBody;
        TextView messageText = (TextView)findViewById(R.id.messageText);
        messageText.setText("Touch NFC Tag to share e-mail\n" +
                    "Mail addres: " + mailAddress+
                    "\nSubject: " + mailSubject + "\nBody: " + mailBody);
    }
});
```

The onNewIntent() method writes e-mail to the NFC tag in the same way as writing the URL and SMS codes—the only difference is in the prefix. The first byte of the payload byte array is created as 0x06 for the mailto: prefix. The following code shows the complete onNewIntent() method:

```
@Override
public void onNewIntent(Intent intent) {
    Log.i("Foreground dispatch", "Discovered tag with intent: " + intent);
    Tag tag = intent.getParcelableExtra(NfcAdapter.EXTRA_TAG);
    byte[] uriField = urlAddress.getBytes(Charset.forName("US-ASCII"));
    byte[] payload = new byte[uriField.length + 1];
    payload[0] = 0x06;
    System.arraycopy(uriField, 0, payload, 1, uriField.length);
    NdefRecord URIRecord  = new NdefRecord(
    NdefRecord.TNF_WELL_KNOWN, NdefRecord.RTD_URI, new byte[0], payload);
    NdefMessage newMessage= new NdefMessage(new NdefRecord[] { URIRecord });
    writeNdefMessageToTag(newMessage, tag);
}
```

Android Manifest

The Android manifest file does not include any NFC tag intent filters, because intent filtering is done with a foreground dispatch system as described in Chapter 5. It includes an NFC permission to use the NFC adapter and activities in order to run those activities. The manifest file is given in Listing 6-1.

LISTING 6-1: NFC Poster Tag Writer Manifest File (NFCPosterTagWriter\AndroidManifest.xml)

```xml
<?xml version="1.0" encoding="utf-8"?>
<manifest xmlns:android="http://schemas.android.com/apk/res/android"
 package="com.nfclab.nfcpostertagwriter"
 android:versionCode="1"
 android:versionName="1.0" >
    <uses-sdk android:minSdkVersion="14" />
    <uses-permission android:name="android.permission.NFC" />

    <application
     android:icon="@drawable/ic_launcher"
     android:label="@string/app_name" >
        <activity
         android:name=".NFCPosterWriterActivity"
         android:label="@string/app_name" >
            <intent-filter>
                <action android:name="android.intent.action.MAIN" />
                <category android:name="android.intent.category.LAUNCHER" />
            </intent-filter>
        </activity>

        <activity
         android:name="com.nfclab.nfcpostertagwriter.WriteUrlActivity">
        </activity>

        <activity
         android:name="com.nfclab.nfcpostertagwriter.WritePhoneActivity">
        </activity>

        <activity
         android:name="com.nfclab.nfcpostertagwriter.WriteSmsActivity">
        </activity>

        <activity
         android:name="com.nfclab.nfcpostertagwriter.WriteMapActivity">
        </activity>

        <activity
         android:name="com.nfclab.nfcpostertagwriter.WriteMailActivity">
        </activity>
    </application>

</manifest>
```

Smart Poster Reader Application

The Smart Poster tag writer application demonstrated how NFC tags for smart posters are encoded. Tag writing applications are used by the system developers before firing the projects; in fact, the users of the projects will need the application that read the tags, not write to them. This section explains the application that will be used by the end users who actually use the smart posters. The application will be able to read any one of the five different data types that are already encoded by the writer application as described previously.

The initial screen of the application is shown in Figure 6-19. Since the application will mostly be launched after NFC interaction, the main screen does not contain much information. It only tells the user to touch to an NFC tag if it is manually launched. Of course, you can create a fancier main screen in your own applications.

The activities run by NFC interaction are as follows:

➤ UrlActivity

➤ PhoneActivity

➤ SmsActivity

➤ MapActivity

➤ MailActivity

FIGURE 6-19

Reading a URL from an NFC Tag

When a tag with TNF_WELL_KNOWN with RTD_URI data is discovered and its scheme is found to be http, the application launches the UrlActivity class automatically and the screen shown in Figure 6-20 is displayed on the mobile. When the "Open Link" button is clicked, the website is opened in the default browser (see Figure 6-21). In order for the UrlActivity class to run automatically, an intent filter in the related activity needs to be defined. See Figure 6-22 for the activity flow diagram of the URLActivity.

FIGURE 6-20   **FIGURE 6-21**

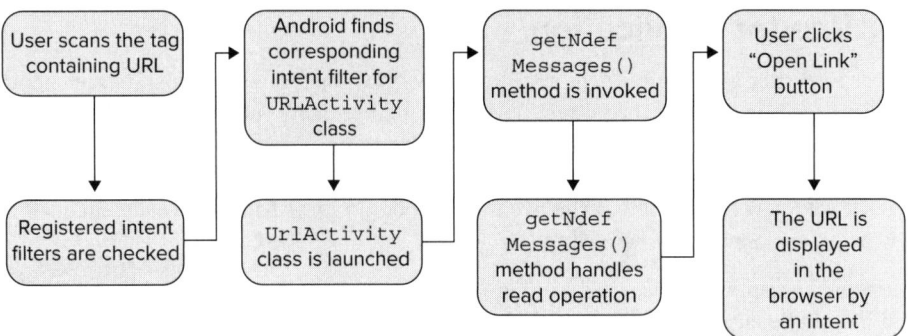

FIGURE 6-22

Intent Filter in Android Manifest when Reading a URL

Remember that the intent filter in the manifest file enables automatic execution of the related activity. Since the URL tags are encoded with `TNF_WELL_KNOWN` with `RTD_URI`, the following intent filter needs to be defined in the manifest file. The `data` element in the intent filter specifies that all URLs with the `http` scheme are filtered and when a related tag is discovered, the `UrlActivity` class is called:

```
<activity android:name=".UrlActivity">
    <intent-filter>
        <action android:name="android.nfc.action.NDEF_DISCOVERED"/>
        <category android:name="android.intent.category.DEFAULT"/>
        <data android:scheme="http"
         android:host="*"
         android:pathPrefix="" />
    </intent-filter>
</activity>
```

UrlActivity Class

Handling `TNF_WELL_KNOWN` with `RTD_URI` was described in Chapter 5. First, the intent is checked to determine if there is an NFC interaction of `NDEF_DISCOVERED`. Then the `getNdefMessages()` method is called, which will handle all the operations and return an `NdefMessage` array as a result. Then, the payload and payload header are extracted from the NDEF record to process the data in the tag:

```
if (NfcAdapter.ACTION_NDEF_DISCOVERED.equals(getIntent().getAction()))
{
    NdefMessage[] messages = getNdefMessages(getIntent());
    for(int i = 0; i<messages.length; i++)
    {
        for(int j = 0; j<messages[0].getRecords().length; j++)
        {
            NdefRecord record = messages[i].getRecords()[j];
            payload = new String(record.getPayload(),1,
                                 record.getPayload().length-1,
                                 Charset.forName("UTF-8"));
            messageText2.setText( payload );
            payloadHeader = record.getPayload()[0];
        }
    }
}
```

When the "Open Link" button is clicked, the `onClickListener()` method is invoked as shown in the following code. The method checks the payload header again and if the header is `0x01` (`http://www.`), a new intent is created to open the website:

```
urlButton.setOnClickListener(new android.view.View.OnClickListener() {
    public void onClick(View v) {
        if(payloadHeader == 0x01)
        {
            Intent data = new Intent();
            data.setAction(Intent.ACTION_VIEW);
            data.setData(Uri.parse("http://www."+payload));
            try {
                startActivity(data);
            } catch (ActivityNotFoundException e) {
                return;
            }
        }
    }
});
```

Reading a Phone Number from an NFC Tag

When `RTD_URI` data is discovered in the tag and its scheme is `tel`, the `PhoneActivity` class is launched and the screen shown in Figure 6-23 is displayed. When the "Call Number" button is clicked, the phone number is processed with an intent for the mobile phone to call (see Figure 6-24). The intent filter defined in the manifest file enables the `PhoneActivity` class to launch automatically when the `tel` scheme is discovered in the tag. See Figure 6-25 for the activity flow diagram of the `PhoneActivity`.

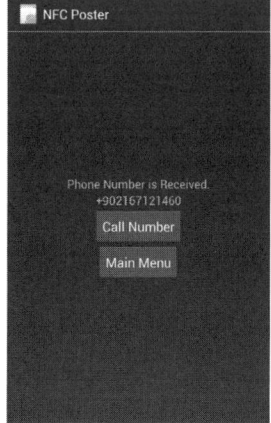

FIGURE 6-23   **FIGURE 6-24**

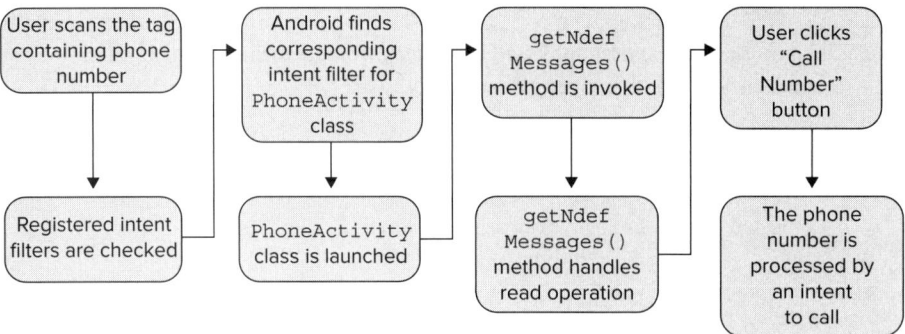

FIGURE 6-25

Intent Filter in Android Manifest when Reading a Phone Number

Since the tags for phone numbers are encoded with TNF_WELL_KNOWN with RTD_URI data type as described in previous sections, the following intent filter needs to be defined in the manifest file in order to filter the URIs with the tel scheme. In this way, as a corresponding tag is discovered, the PhoneActivity class is invoked:

```
<activity android:name=".PhoneActivity">
    <intent-filter>
        <action android:name="android.nfc.action.NDEF_DISCOVERED"/>
        <category android:name="android.intent.category.DEFAULT"/>
        <data android:scheme="tel" />
    </intent-filter>
</activity>
```

PhoneActivity Class

Earlier in this chapter, you used the same data type as used for URLs (TNF_WELL_KNOWN with RTD_URI) in order to filter phone numbers in the tags. So, most of the code for the PhoneActivity class is similar to the code for the UrlActivity class. The difference is in the onClickListener() method, which creates the new intent when the "Call Number" button is clicked to dial the discovered phone number:

```
phoneButton.setOnClickListener(new android.view.View.OnClickListener() {
    public void onClick(View v) {
        if(payloadHeader == 0x05)
        {
            Intent data = new Intent();
            data.setAction(Intent.ACTION_DIAL);
            data.setData(Uri.parse( "tel://" + payload ));
            try {
                startActivity(data);
            } catch (ActivityNotFoundException e) {
                return;
            }
        }
    }
});
```

Reading an SMS from an NFC Tag

When a `TNF_EXTERNAL_TYPE` record with the external type of `nfclab.com:smsService` is discovered in the tag, the `SmsActivity` class is invoked. When the activity is first executed, the screen shown in Figure 6-26 is displayed. If the user clicks the "Send SMS" button, the Android SMS application is run by an intent and the received data are passed to the SMS application. This time the user sees a screen similar to the one in Figure 6-27. See Figure 6-28 for the activity flow diagram of the `SmsActivity`.

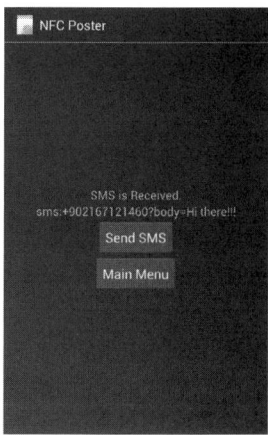

FIGURE 6-26

FIGURE 6-27

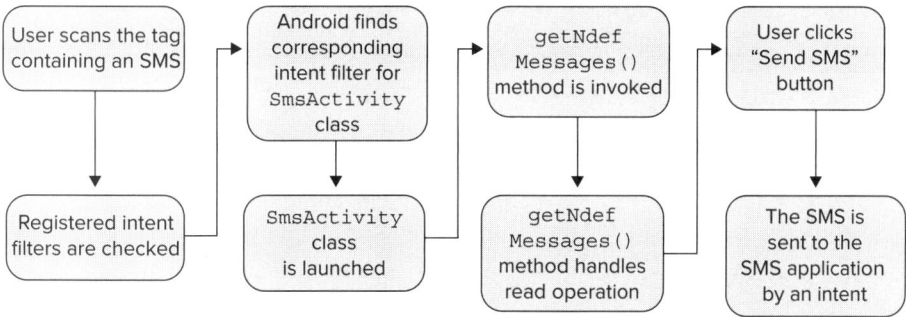

FIGURE 6-28

Intent Filter in Android Manifest when Reading an SMS

Remember that the tags for SMS messages are coded with `TNF_EXTERNAL_TYPE`. So, the following intent filter needs to be defined in the manifest file to invoke the activity when the external type of `nfclab.com:smsService` is discovered inside the NDEF record:

```
<activity android:name=".SmsActivity">
    <intent-filter>
    <action android:name="android.nfc.action.NDEF_DISCOVERED" />
    <category android:name="android.intent.category.DEFAULT" />
```

```
     <data android:scheme="vnd.android.nfc"
      android:host="ext"
      android:pathPrefix="/nfclab.com:smsService"/>
   </intent-filter>
</activity>
```

SmsActivity Class

Reading tags with `TNF_EXTERNAL_TYPE` was described in Chapter 5. The following code processes the NDEF message and receives the payload data:

```
if (NfcAdapter.ACTION_NDEF_DISCOVERED.equals(getIntent().getAction()))
{
   NdefMessage[] messages = getNdefMessages(getIntent());
   for(int i=0; i<messages.length; i++)
   {
     for(int j=0;j<messages[0].getRecords().length; j++)
     {
       NdefRecord record = messages[i].getRecords()[j];
       payload = new String(record.getPayload());
       messageText2.setText( payload );
       payloadHeader = record.getPayload()[0];
     }
   }
}
```

In order to send SMS with the payload, intents need to be used as shown in the following code. The `onClickListener()` method creates the new intent when the "Send SMS" button is clicked to launch the Android SMS application:

```
smsButton.setOnClickListener(new android.view.View.OnClickListener() {
   public void onClick(View v) {
       Intent data = new Intent();
       data.setAction(Intent.ACTION_VIEW);
       data.setData(Uri.parse(payload));
       try {
          startActivity(data);
       } catch (ActivityNotFoundException e) {
          return;
       }
   }
});
```

Reading a Geolocation from an NFC Tag

When a `TNF_EXTERNAL_TYPE` record with the external type of `nfclab.com:geoService` is discovered in the tag, the `MapActivity` class is launched. When the activity is first run, the screen shown in Figure 6-29 is displayed. If the user clicks the "Display on Map" button, the default map application of Android is launched by an intent and the discovered data are passed to the application in order to show the location on the map (see Figure 6-30). The manifest file with the intent filter enables the `MapActivity` class to launch automatically when the external type of the `nfclab.com:geoService` scheme is discovered. See Figure 6-31 for the activity flow diagram of the `MapActivity`.

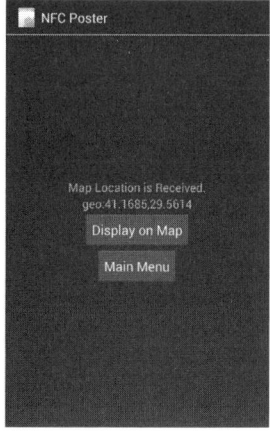

FIGURE 6-29

FIGURE 6-30

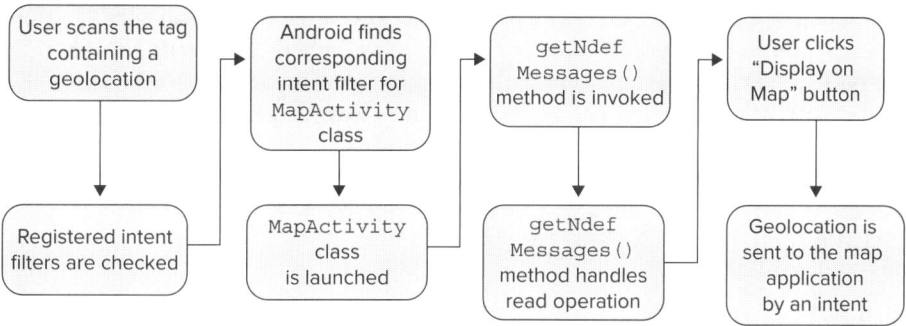

FIGURE 6-31

Intent Filter in Android Manifest when Reading a Geolocation

The tags for geolocations are coded with TNF_EXTERNAL_TYPE as described earlier. Therefore, the following intent filter needs to be defined in the manifest file to invoke the MapActivity class when the external type of nfclab.com:geoService is discovered inside the NDEF record:

```
<activity android:name=".MapActivity">
    <intent-filter>
        <action android:name="android.nfc.action.NDEF_DISCOVERED" />
        <category android:name="android.intent.category.DEFAULT" />
        <data android:scheme="vnd.android.nfc"
         android:host="ext"
         android:pathPrefix="/nfclab.com:geoService"/>
    </intent-filter>
</activity>
```

MapActivity Class

Earlier in this chapter, you used the same data type as used for SMS messages (TNF_EXTERNAL_TYPE) in order to filter geolocations in the tags. Thus, most of the code is similar to the SmsActivity class.

As shown in the following code, only some changes are made in the `onClickListener()` method to create a new intent to display the location on a map when the "Display on Map" button is clicked:

```
mapButton.setOnClickListener(new android.view.View.OnClickListener() {
    public void onClick(View v) {
        Intent data = new Intent();
        data.setData(Uri.parse( payload ));
        data.setAction(Intent.ACTION_VIEW);
        try {
            startActivity(data);
        } catch (ActivityNotFoundException e) {
            return;
        }
    }
});
```

Reading an E-mail from an NFC Tag

When an `RTD_URI` is discovered with a data scheme of `mailto`, the `MailActivity` class is invoked. Then, the screen shown in Figure 6-32 is activated. When the user clicks the "Send Mail" button, the e-mail data are passed to the default e-mail application (see Figure 6-33). Again, the manifest file with the intent filter enables the `MailActivity` class to launch automatically when the `mailto` scheme is discovered in the tag. See Figure 6-34 for the activity flow diagram of the `MapActivity`.

FIGURE 6-32   **FIGURE 6-33**

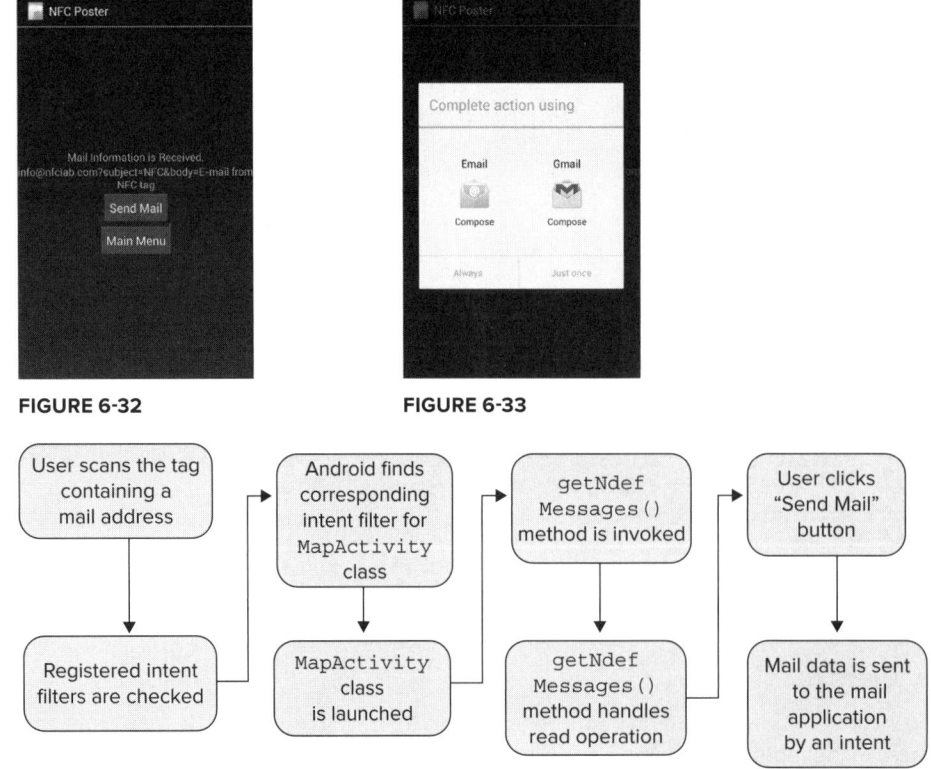

FIGURE 6-34

Intent Filter in Android Manifest when Reading an E-mail

The tags encoded for e-mails hold the data of `TNF_WELL_KNOWN` with `RTD_URI`. Also the following intent filter should be defined in the manifest file. In this way, when a tag with an NDEF message is discovered and the first NDEF record's URI scheme is `mailto`, the `MailActivity` class is run automatically:

```
<activity
   android:name=".MailActivity">
   <intent-filter>
      <action android:name="android.nfc.action.NDEF_DISCOVERED"/>
      <category android:name="android.intent.category.DEFAULT"/>
      <data android:scheme="mailto" />
   </intent-filter>
</activity>
```

MailActivity Class

Earlier in this chapter, you used the same data type with the `UrlActivity` and `PhoneActivity` classes (`TNF_WELL_KNOWN` with `RTD_URI`) in order to filter e-mails in the tags. So, most of the code for the `MailActivity` class is similar to the code for the `UrlActivity` and `PhoneActivity` classes. As shown in the following code, you only need to make some changes in the `onClickListener()` method to create a new intent to send e-mail when the "Send Mail" button is clicked (see Figure 6-34):

```
mailButton.setOnClickListener(new android.view.View.OnClickListener() {
    public void onClick(View v) {
        if( payloadHeader == 0x06)
        {
            Intent data = new Intent();
            data.setAction(Intent.ACTION_VIEW);
            data.setData(Uri.parse("mailto:" + payload));
            try {
                startActivity(data);
            } catch (ActivityNotFoundException e) {
                return;
            }
        }
    }
});
```

NFC SHOPPING USE CASE

The NFC shopping use case enables remote shopping by touching tags of the products. Tags can be placed onto the package of each product or on fliers promoting products.

When the application is launched, the main screen directs the user to touch to an NFC tag to read the product data (see Figure 6-35). As the mobile is touched to the tag, the product data is transferred to the mobile and the user is asked to specify a quantity (see Figure 6-36). After the user inputs the quantity and clicks the "Add to Basket" button, the product, together with the requested quantity, is added to the basket.

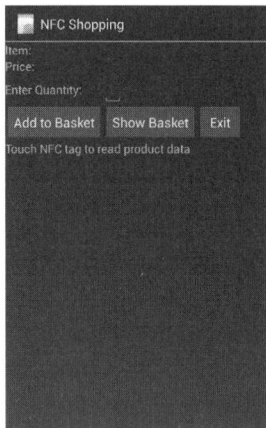

FIGURE 6-35

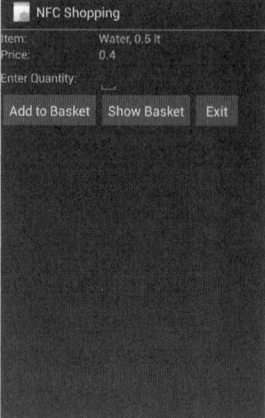

FIGURE 6-36

When the "Show Basket" button is clicked, all products in the basket are listed on the screen (see Figure 6-37). After all the required products have been added to the basket, the user can finalize their shopping and order the contents of the basket by clicking the "Order Basket" button.

The data in the tags are encoded with a `TNF_EXTERNAL_TYPE` record with the external type of `nfclab.com:shopping`. Both the writer and reader applications are programmed as described in the following sections.

NFC Shopping Tag Writer Application

The NFC shopping application is a proprietary application for markets and other shopping centers. Therefore, only the corresponding reader application needs to be able to process the data in tags. When well-known data types are used, many applications can process the data. So, in order to create NDEF records, a `TNF_EXTERNAL_TYPE` record with a special external type, `nfclab.com:shopping`, is used.

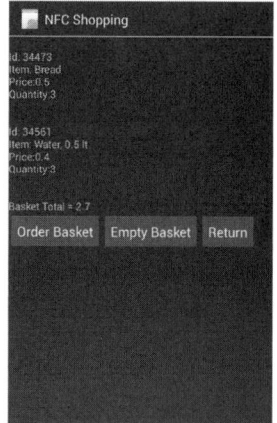

FIGURE 6-37

The main activity of the application includes three `EditText` objects: item ID, item name, and item price (see Figure 6-38). When the fields are filled with the required data and the "Save to Tag" button is clicked, the mobile device waits for an NFC tag to write the NDEF message (see Figure 6-39). When the tag is discovered in proximity, the NDEF message is written to the tag (see Figure 6-40). See Figure 6-41 for the activity flow diagram of the NFC shopping tag writer application.

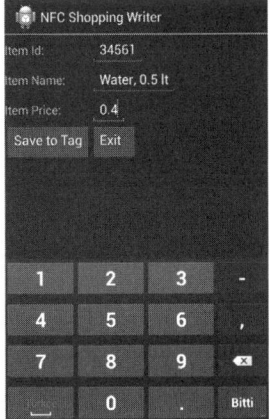

FIGURE 6-38

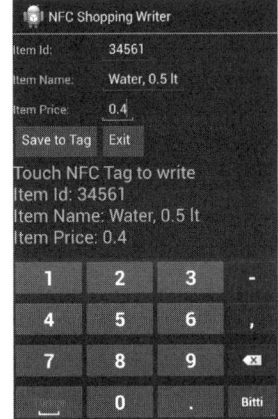

FIGURE 6-39

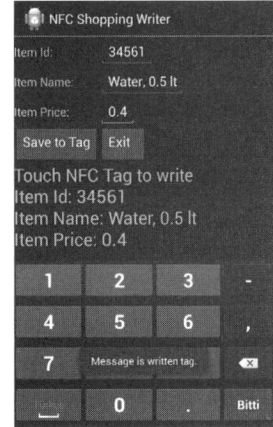

FIGURE 6-40

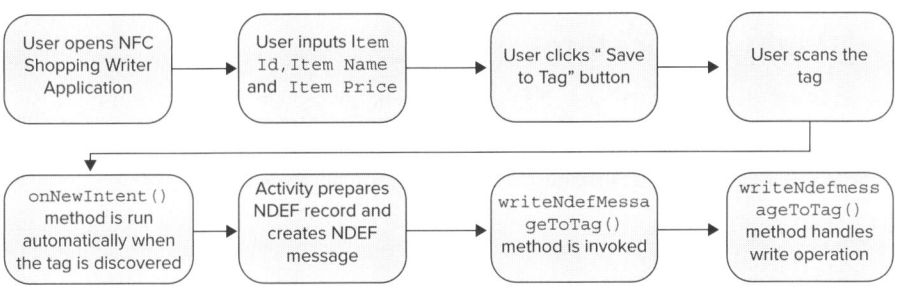

FIGURE 6-41

When the "Save to Tag" button is clicked, the `onClickListener()` method of the activity is invoked. In this method, the values in the `EditText` object are saved to three different strings, and a message is displayed instructing the user to touch to an NFC tag as shown in the following code:

```
writeItemButton.setOnClickListener(new View.OnClickListener() {
    public void onClick(View view) {
        itemId = itemIdEditText.getText().toString();
        itemName = itemNameEditText.getText().toString();
        itemPrice = itemPriceEditText.getText().toString();
        TextView messageText = (TextView)findViewById(R.id.messageText);
        messageText.setText("Touch NFC Tag to write \n");
        messageText.append("Item Id: "+ itemId + "\n" +
                    "Item Name: " + itemName + "\nItem Price: " + itemPrice);
    }
});
```

The application is implemented with the foreground tag dispatch system. As explained in the "Foreground Dispatch System" section in Chapter 5, when a desired tag is discovered in proximity, the `onNewIntent()` method is invoked. In order to save three different values to one NDEF record, these values are merged to one string named `payload`, and they are delimited with a colon in order to split them easily after reading the tag backwards. As an alternative, you may save each value to a separate NDEF record; however, this time the size of the NDEF message will grow unnecessarily. After creating the payload string, an NDEF record is created with the `TNF_EXTERNAL_TYPE` record of an `nfclab.com:shopping` external type. Finally, the tag and the NDEF message are sent to the `writeNdefMessageToTag()` method, which writes the NDEF message to the tag. The related code part is shown here:

```
@Override
public void onNewIntent(Intent intent) {
    Log.i("Foreground dispatch", "Discovered tag with intent: " + intent);
    Tag tag = intent.getParcelableExtra(NfcAdapter.EXTRA_TAG);
    String externalType = "nfclab.com:shopping";
    String payload = itemId+":"+itemName+":"+itemPrice;
    NdefRecord extRecord1 = new NdefRecord(NdefRecord.TNF_EXTERNAL_TYPE,
                              externalType.getBytes(), new byte[0],
                              payload.getBytes());
    NdefMessage newMessage = new NdefMessage(new NdefRecord[] { extRecord1});
    writeNdefMessageToTag(newMessage, tag);
}
```

NFC Shopping Main Application

The tag writer application demonstrated how NFC tags are encoded for the NFC shopping case. This section describes how the main application will be used by the end users.

Five classes are created for the application, and three of them are activities:

➤   `NFCShoppingActivity.java`

➤   `ShowBasketActivity.java`

➤   `OrderActivity.java`

➤   `Item.java`

➤   `Basket.java`

Manifest File

The manifest file includes NFC permission, `intent-filter` to filter corresponding NFC tags, and activities. Since the tags are encoded with the `TNF_EXTERNAL_TYPE` record with the external type of `nfclab.com:shopping`, the related intent filter is defined in the manifest file for the `NFCShoppingActivity` class. Listing 6-2 gives the content of the manifest file.

LISTING 6-2: NFC Shopping Tag Writer Manifest File (NFCShoppingTagWriter\AndroidManifest.xml)

```
<?xml version="1.0" encoding="utf-8"?>
<manifest xmlns:android="http://schemas.android.com/apk/res/android"
 package="com.nfclab.nfcshopping"
 android:versionCode="1"
 android:versionName="1.0" >
    <uses-sdk android:minSdkVersion="14" />
```

```
<uses-permission android:name="android.permission.NFC" />

<application
 android:icon="@drawable/ic_launcher"
 android:label="@string/app_name_shopping" >
    <activity
     android:name=".NFCShoppingActivity"
     android:label="@string/app_name_shopping" >
        <intent-filter>
            <action android:name="android.intent.action.MAIN" />
            <category android:name="android.intent.category.LAUNCHER" />
        </intent-filter>
        <intent-filter>
            <action android:name="android.nfc.action.NDEF_DISCOVERED" />
            <category android:name="android.intent.category.DEFAULT" />
            <data android:scheme="vnd.android.nfc"
             android:host="ext"
             android:pathPrefix="/nfclab.com:shopping"/>
        </intent-filter>
    </activity>
    <activity android:name=".ShowBasketActivity"> </activity>
    <activity android:name=".OrderActivity"> </activity>
</application>

</manifest>
```

NFCShoppingActivity Class

When the mobile is touched to the corresponding NFC tag, the following code is run:

```
if (NfcAdapter.ACTION_NDEF_DISCOVERED.equals(getIntent().getAction()))
{
    NdefMessage[] messages = getNdefMessages(getIntent());
    for(int i=0; i<messages.length; i++)
    {
        for(int j = 0; j<messages[0].getRecords().length; j++)
        {
            NdefRecord record = messages[i].getRecords()[j];
            payload = new String(record.getPayload(), 0, record.getPayload().length,
                           Charset.forName("UTF-8"));
            String delimiter = ":";
            String[] temp = payload.split(delimiter);
            itemId = temp[0];
            itemName = temp[1];
            itemPrice = temp[2];
            nameTextView.setText( itemName );
            priceTextView.setText( itemPrice );
            messageText.setText("");
        }
    }
}
```

First, the intent is checked to see if there is an NFC interaction of NDEF_DISCOVERED. Then the getNdefMessages() method is called to receive the NDEF message from the tag and the payload and payload header are extracted from the NDEF record to process the data in the tag. Afterwards, the payload needs to be processed further, because there are three different data (item ID, item name, and item price) in the tag, separated by a colon (:) delimiter. So, the related method is called to split the payload to gather three different data, which are item ID, item name, and item price. Now all the user needs to do is enter the quantity and add the product to the basket. When the user clicks the "Add to Basket" button, the `Basket.add(name, price, quantity)` method is invoked. Please see Figure 6-42 for the activity flow diagram of the NFC shopping application.

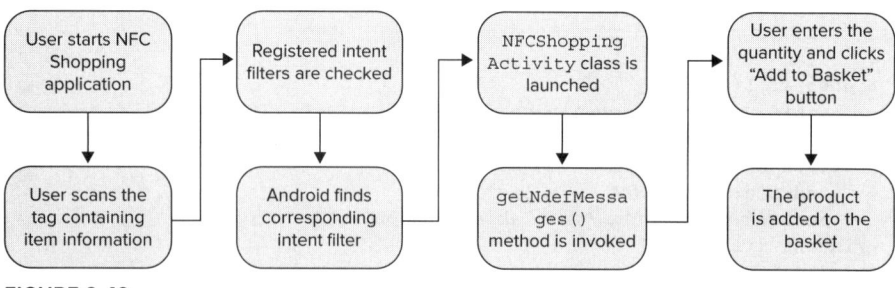

FIGURE 6-42

ShowBasketActivity.java

The `ShowBasketActivity` class consists mainly of a `TextView` item named `items` and three buttons: `orderButton`, `emptyButton`, and `exitButton`. `items` is used to display the items within the basket; `orderButton` is used to trigger ordering the items of the basket; `emptyButton` is used to trigger clearing the content of the basket; and `exitButton` is used to return to the previous activity.

When `orderButton` is clicked, the following code will create an instance of the `OrderActivity` class, and cause the `onCreate()` method of the `OrderActivity` class to be executed:

```
orderButton.setOnClickListener(new android.view.View.OnClickListener() {
    public void onClick(View view) {
        Intent myIntent = new Intent( view.getContext(), OrderActivity.class);
        startActivityForResult(myIntent, 0);
    }
});
```

When `emptyButton` is clicked, the `Basket.empty()` method will be executed, which will immediately clear the contents of the basket. When `exitButton` is clicked, the `finish()` method will be invoked.

OrderActivity.java

Remember that the `OrderActivity` class is triggered when `orderButton` in the `ShowBasketActivity` class is clicked. The activity should send the order in the basket to the market or shopping center; however, the implementation of this is beyond the scope of this book and is left to the user.

Item.java

The `Item` class is used to define the item to be manipulated by the `Basket` class. This contains `id`, `name`, `price`, and the `quantity` as the class attributes. The class also contains the required getter and setter methods.

Basket.java

The `Basket` class manipulates the basket. It mainly contains an `add()` method to add an item, a `total()` method to calculate the cost of the items in the basket, and an `empty()` method to empty the basket.

STUDENT TRANSPORTATION TRACKING USE CASE

This use case enables tracking the students' transportation activity from home to school and from school to home. In order to run the application, each student needs to have an NFC tag that will be encoded to contain the ID of that student. It is assumed that, inside the school bus, there is an NFC-enabled mobile phone, possibly owned by the driver that reads the tags of the students. When a student gets onto the bus, they touch their tag to the mobile and the mobile saves the getting-on time (see Figure 6-43); when that student gets off from the bus, they again touch their tag to the mobile and this time the mobile saves the getting-off time (see Figure 6-44).

FIGURE 6-43

FIGURE 6-44

The application can also be extended to send the getting-on and getting-off times to a server online, so that the students' parents can see the information via a website or mobile application that is integrated to the system.

The data in the tags are encoded with a `TNF_EXTERNAL_TYPE` record with the external type of `nfclab.com:transport`. Both the writer and reader applications are described in the following sections.

Student Tracking Tag Writer Application

The student tracking application is a proprietary application that is in some ways similar to NFC shopping. Therefore, a `TNF_EXTERNAL_TYPE` record with a special external type is used.

The writer application includes an activity to write the students' data to tags. Two `EditText` objects are included for the student ID and student name. When a student's data are inputted and the "Save to Tag" button is clicked, the `onClickListener()` method is invoked. Inputted values are saved to two different strings and a message is displayed to the user that directs them to touch to an NFC tag as shown in the following code:

```
writeStudentButton.setOnClickListener(new android.view.View.OnClickListener() {
    public void onClick(View view) {
        studentId = studentIdEditText.getText().toString();
        studentName = studentNameEditText.getText().toString();
        TextView messageText = (TextView)findViewById(R.id.messageText);
        messageText.setText("Touch NFC Tag to write \n");
        messageText.append("Student id:" + studentId +
                        "\nStudent Name: " + studentName );
    }
});
```

When the tag is discovered in proximity, the prepared NDEF message is written to the tag (see Figure 6-45). Please see Figure 6-46 for the activity flow diagram of the tag writer application.

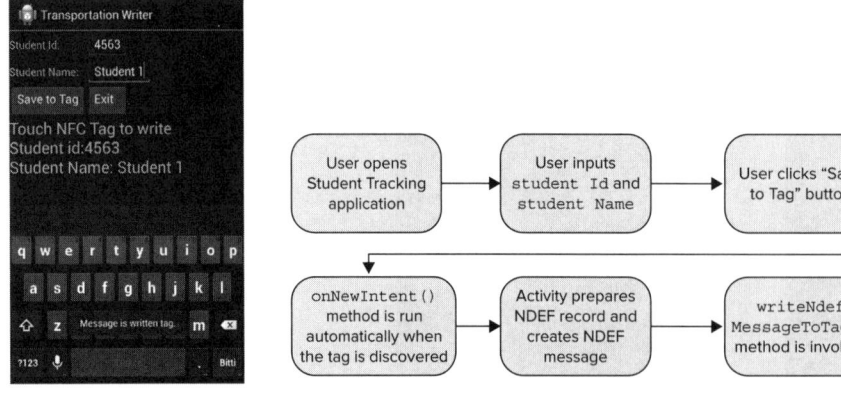

FIGURE 6-45   **FIGURE 6-46**

In the `onNewIntent()` method, the student ID and student name are merged to form a string named `payload` and separated with a colon. The colon will be used as a delimiter to split them in the reading stage. Then a new NDEF record is prepared with the payload and the external type of `nfclab.com:transport`. Since there is only one NDEF record to be saved to the tag, a new NDEF message is created from this NDEF record and the `writeNdefMessageToTag()` method is invoked using the necessary parameters to write the NDEF message to the tag. The related code is shown here:

```
@Override
public void onNewIntent(Intent intent) {
    Log.i("Foreground dispatch", "Discovered tag with intent: " + intent);
    Tag tag = intent.getParcelableExtra(NfcAdapter.EXTRA_TAG);
    String externalType = "nfclab.com:transport";
    String payload = studentId + ":" + studentName;
    NdefRecord extRecord1 = new NdefRecord(NdefRecord.TNF_EXTERNAL_TYPE,
                                    externalType.getBytes(), new byte[0],
                                    payload.getBytes());
    NdefMessage newMessage = new NdefMessage(new NdefRecord[] { extRecord1});
    writeNdefMessageToTag(newMessage, tag);
}
```

Student Tracking Main Application

Within the application, a manifest file and the following four Java classes are created:

➤ `TransportationActivity.java`

➤ `WebServiceActivity.java`

➤ `Group.java`

➤ `Student.java`

Manifest File

The manifest file includes the activities, permission to use NFC adapter, and `intent-filter` to filter NFC tags with an `nfclab.com:transport` external type. The content of the manifest file is given in Listing 6-3.

LISTING 6-3: NFC Shopping Tag Writer Manifest File (NFCStudentTrackingTagWriter\
AndroidManifest.xml)

```
<?xml version="1.0" encoding="utf-8"?>
<manifest xmlns:android="http://schemas.android.com/apk/res/android"
 package="com.nfclab.transportation"
 android:versionCode="1"
 android:versionName="1.0" >

    <uses-sdk android:minSdkVersion="14" />
    <uses-permission android:name="android.permission.NFC" />

    <application
     android:icon="@drawable/ic_launcher"
     android:label="@string/app_name_transportation" >
        <activity
         android:name=".TransportationActivity"
         android:label="@string/app_name_transportation" >
            <intent-filter>
                <action android:name="android.intent.action.MAIN" />
                <category android:name="android.intent.category.LAUNCHER" />
            </intent-filter>
            <intent-filter>
```

continues

LISTING 6-3 *(continued)*

```
                    <action android:name="android.nfc.action.NDEF_DISCOVERED" />
                    <category android:name="android.intent.category.DEFAULT" />
                    <data android:scheme="vnd.android.nfc"
                     android:host="ext"
                     android:pathPrefix="/nfclab.com:transport"/>
                </intent-filter>
            </activity>
            <activity android:name=".WebServiceActivity"> </activity>
        </application>

    </manifest>
```

TransportationActivity Class

When a student touches their tag to the mobile phone, the data in the tag is transferred to the mobile. After the data is read, the string is split into two parts by identifying the colon, and then the student ID and student name are saved to two different strings as shown in the following code:

```
if (NfcAdapter.ACTION_NDEF_DISCOVERED.equals(getIntent().getAction()))
{
    NdefMessage[] messages = getNdefMessages(getIntent());
    for(int i=0; i<messages.length; i++)
    {
        for(int j=0; j<messages[0].getRecords().length; j++)
        {
            NdefRecord record = messages[i].getRecords()[j];
            payload = new String(record.getPayload());
            String delimiter = ":";
            String[] temp = payload.split(delimiter);
            studentId = temp[0];
            studentName = temp[1];
        }
    }
}
```

Remember also here that the application on the mobile phone keeps a list of the students on the transportation service. If the student is getting on the service, the application adds the student information to the list together with their getting-on time. When the same student gets off the service, the mobile reads the student information from the tag for the second time, then finds that student's information in the existing student list, and updates the getting-off time. Please see Figure 6-47 for the activity flow diagram of the application.

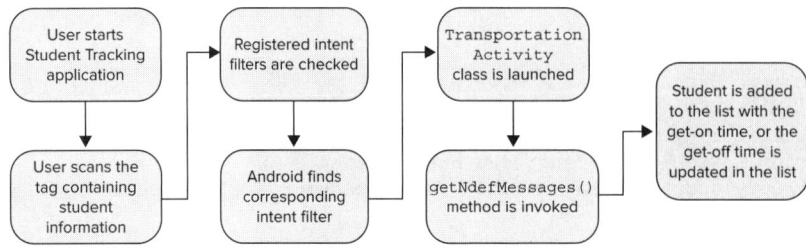

FIGURE 6-47

WebServiceActivity.java

The web service activity aims to upload the activity information of the students to a web server. This way, parents who have the appropriate web or mobile interface will be able to track the activities of their children. Writing source code for this web service activity is beyond the scope of this book, but a competent software developer can easily create this code.

Group.java

This class basically maintains the list of students who are on the service bus, specifically their getting-on and getting-off activities. It internally uses an `ArrayList`, called `students`. It reads the student data that are processed and searches for the student in the currently existing list. If the student is not on the list, the application adds the student information to the list, together with their getting-on time. If the student is already in the list, the getting-off time is updated in the student's record.

Student.java

This class consists of the `Student` type declaration and its setter and getter methods. It is instantiated by `Group.java`.

SUMMARY

In this chapter, you saw three different use cases in reader/writer mode. After reading this chapter, you can implement many more additional cases, and many innovative ways to use this mode.

In the Smart Poster use case, the webpage, e-mail, and phone number scenarios are implemented with `TNF_WELL_KNOWN` with the `RTD_URI` record type. However, the SMS and geolocation scenarios are implemented with the `TNF_EXTERNAL_TYPE` record type, because NFC Forum has not yet defined any URI identifier code for those URIs.

The second use case is an NFC-based remote shopping application that enables mobile shopping using NFC technology. When the mobile phone is touched to an NFC tag, the corresponding product is added to the basket. The case is implemented using the `TNF_EXTERNAL_TYPE` record type since the application needs to be a proprietary application and other applications should not handle the corresponding tags.

The third use case tracks students' transportation activity between school and home and vice versa. When a student gets on or off the bus, they touch their NFC tag to the mobile phone in the bus and their transportation status is updated. The case is implemented with the `TNF_EXTERNAL_TYPE` record type since the application needs to be a proprietary application like the shopping application.

7

NFC Programming: Peer-to-Peer Mode

WHAT'S IN THIS CHAPTER?

- ➤ Introduction to peer-to-peer mode programming

- ➤ How to perform a peer-to-peer transaction

- ➤ Beaming NDEF messages with the `setNdefPushMessageCallback()` method

- ➤ Beaming NDEF messages with the `setNdefPushMessage()` method

- ➤ Beaming NDEF messages with the `enableForegroundNdefPush()` method

- ➤ Receiving beams

WROX.COM CODE DOWNLOADS FOR THIS CHAPTER

The wrox.com code downloads for this chapter are found at `www.wrox.com/remtitle` `.cgi?isbn=1118380096` on the Download Code tab. The code is in the Chapter 7 download and individually named according to the names throughout the chapter.

In this chapter, peer-to-peer mode application programming is demonstrated. In Android OS, sending data from one mobile to another via peer-to-peer mode is called a beam. An application that beams can be implemented in two different ways: using either the `setNdefPushMessageCallback()` method or the `setNdefPushMessage()` method. This chapter describes and gives examples of the properties and implementation of both of these. It also describes the implementation of receiving beams, which is the same in both methods.

These two methods can be used in devices that have an API level of 14 or higher. Older devices that have an API level of 10 to 13 can use foreground NDEF Push to beam messages as well. You should keep in mind that this methodology is deprecated starting from API level 14, hence you shouldn't use it if you don't have to. Moreover, in earlier versions of API level 9 and lower, sending beams using peer-to-peer mode is not possible.

Peer-to-peer mode has some common usages with reader/writer mode. These usages are also described in this chapter.

PERFORMING PEER-TO-PEER TRANSACTIONS

As described in Chapter 1, "Overview of Near Field Communication," peer-to-peer mode allows two mobile phones to exchange data when they touch each other. In order to beam between two mobile devices, the following preconditions must be met:

➤ Android Beam from the settings of Android OS must be turned on in both mobile phones.

➤ The application on the mobile phone that wants to beam must be running in the foreground.

➤ The screen of the mobile phone that will receive the beam must not be locked.

➤ When two mobile phones touch each other, the Touch to Beam UI is displayed in both mobile phones (see Figure 7-1). The user must touch to the screen of the mobile phone that will beam.

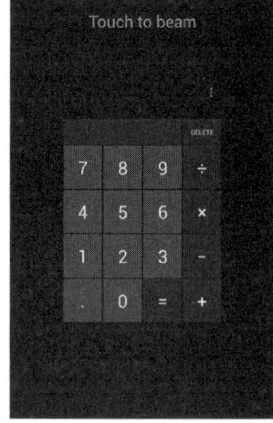

FIGURE 7-1

In Android, you can beam NDEF messages to other mobile devices. In order to beam an NDEF message, you should first create and prepare it as described in Chapter 5, "NFC Programming: Reader/Writer Mode." It is important to note that if your activity does not implement Android Beam, the activity will still beam to the target device but this time the activity will beam a default NDEF message, which is the URI of your application. If the target device's Android API level is between 9 and 13, the device will notify the user that the target application is not present. If its API level is 14 or higher, then the device will try to locate the application in Google Play in order to download and install it. With this functionality, users may also share the application with other users without the need to search for it on Google Play, which is valid for API level 14 or higher. Let's assume that you have opened Google Maps from your device and touched another mobile. The Touch to Beam UI will be displayed on your mobile's screen. If you touch to the screen of your mobile, the target mobile automatically opens Google Maps. If Google Maps is not installed on the target mobile, it will open Google Play to download it.

Another important note is that when you run an activity that implements Android Beam, the tag intent dispatch system is disabled if the activity is in the foreground. Thus, Android cannot scan any tag. If you wish to enable tag scanning as well, you must also implement foreground tag dispatching inside the activity. Please see Chapter 5 for the foreground dispatch system.

In order to beam to another mobile, you need to encapsulate whatever data you want to exchange (such as `TNF_MIME_MEDIA`, `TNF_WELL_KNOWN` with `RTD_URI`, an image file, and so on) to an `NdefMessage` object. The mobile device that receives the beam must support Android's NDEF Push Protocol or NFC Forum's Simple NDEF Exchange Protocol (SNEP). Also note that NDEF Push Protocol is Android-specific so it is supported only by Android. However, SNEP is also supported by other mobile operating systems since it is an NFC Forum specification. If your Android mobile device has NFC capability, both protocols are generally supported. NDEF Push Protocol is required for devices up to API level 13 whereas NDEF Push Protocol and Simple NDEF Exchange Protocol are both required on the devices that have an API level of 14 and later.

You can gather the information from your activity if Android Beam is enabled from the settings. Note that if NFC is enabled for a device, it doesn't mean that Android Beam is also enabled. The method `isNdefPushEnabled()` is added in API level 14 and returns true if both NFC and Android Beam features are enabled. Also note that, if Android Beam is disabled for a device, it can still receive beams but cannot send any. Here is an example of the code:

```
if (!mNfcAdapter.isEnabled()) {
    startActivity(new Intent(Settings.ACTION_NFC_SETTINGS));
} else if (!mNfcAdapter.isNdefPushEnabled()) {
    startActivity(new Intent(Settings.ACTION_NFCSHARING_SETTINGS));
}
```

BEAMING NDEF MESSAGES

You have already learned how to create an NDEF message in reader/writer mode. In order to beam messages, you also need to create an NDEF message similarly. Hence, the steps to create NDEF records and NDEF messages are the same as with reader/writer mode.

There are two options that you can use in order to enable Android Beam in your activity: you can implement either the `setNdefPushMessageCallback()` method or the `setNdefPushMessage()` method.

Beaming with setNdefPushMessageCallback( )

In order to beam using the `setNdefPushMessageCallback()` method, you should follow these steps:

1. Implement `CreateNdefMessageCallback` in your activity.

2. Call `setNdefPushMessageCallback(NfcAdapter.CreateNdefMessageCallback callback, Activity activity, Activity ... activities)`. When this method is called, the activity accepts a callback and if a mobile device is discovered to beam the data, the `createNdefMessage(NfcEvent)` method is invoked automatically.

3. Inside the `createNdefMessage(NfcEvent)` method, create the NDEF message and return it. This NDEF message will be beamed to the target device.

The `setNdefPushMessageCallback()` method dynamically generates NDEF messages. You may call this method anywhere in your activity. The preferred option is to call in the activity's `onCreate()` method.

When the target mobile phone is discovered, the `createNdefMessage(NfcEvent)` method is invoked, which returns an NDEF message. The NDEF message that this method returns is sent to the target mobile. During this time, the Touch to Beam UI is displayed on top of your activity, so you should not get any input from the user in this method. You should only create the NDEF message that you will send, and return it.

Beaming with setNdefPushMessage( )

In order to beam using the `setNdefPushMessage()` method, follow these steps:

1. Create an NDEF message.

2. Call `setNdefPushMessage (NdefMessage message, Activity activity, Activity ... activities)`. When this method is called, the activity sets the received `NdefMessage` parameter as an NDEF message to beam. When the target mobile is discovered, it beams this message.

The `setNdefPushMessage()` method normally sets the NDEF message statically because it gets the NDEF message as a parameter. However, based on your activity's implementation, you may prefer to use this method.

As with the `setNdefPushMessageCallback()` method, you can call this method at any time.

Common Notes

`setNdefPushMessageCallback()` has priority over `setNdefPushMessage()`. So if both are implemented in the activity, `setNdefPushMessageCallback()` will take the priority and the activity will use the callback.

If neither of these two methods has been implemented in your activity, then Android OS sends a default NDEF message (the URI of your application), which opens the same application in the target mobile device. If the application is not installed, the target mobile device opens the application page on Google Play if the device's API level is 14 or higher. If the device's API level is between 9 and 13, it notifies the user that the target application is not present.

If a null NDEF message is set to beam, then NDEF Push will be disabled for the activity.

In order to prevent sending default NDEF messages for all activities of your application, you can include the related `meta-data` element inside the application element as follows:

```
<application ...>
  <meta-data
   android:name="android.nfc.disable_beam_default"
   android:value="true" />
</application>
```

RECEIVING BEAMS

In order to receive beams you should follow these steps:

1. Implement the onNewIntent(Intent) method, and inside this method, call the setIntent(Intent) method. Note that the onResume() method will generally be called after the onNewIntent(Intent) method.

2. In the onResume() method, check if the activity is started with a beam and invoke a method to process the received NDEF message.

AN ABSTRACT BEAM WITH SETNDEFPUSHMESSAGECALLBACK( )

In the code snippet below, an abstract example to send and receive Android Beam using the setNdefPushMessageCallback() method is given:

```java
public class PeertoPeerActivity extends Activity
                                implements CreateNdefMessageCallback
{
    @Override
    public void onCreate(Bundle savedInstanceState) {
        super.onCreate(savedInstanceState);
        setContentView(R.layout.main);
        mNfcAdapter = NfcAdapter.getDefaultAdapter(this);
        if (mNfcAdapter == null) {
            finish();
            return;
        }
        mNfcAdapter.setNdefPushMessageCallback(this, this);
    }

    @Override
    public NdefMessage createNdefMessage(NfcEvent event) {
        //You will write codes to create Ndef message here.
        return message;
    }

    @Override
    public void onResume() {
        super.onResume();
        if (NfcAdapter.ACTION_NDEF_DISCOVERED.equals(getIntent().getAction())) {
            processIntent(getIntent());
        }
    }

    @Override
    public void onNewIntent(Intent intent) {
        setIntent(intent);
    }

    void processIntent(Intent intent) {
        //You will write codes to process incoming Ndef message.
    }
}
```

First of all, you should implement `CreateNdefMessageCallback` in your activity and register a callback using the `setNdefPushMessageCallback(this, this)` method. Then, you need to implement the `createNdefMessage(NfcEvent event)` method in order to create the NDEF message and return it. Remember that the `createNdefMessage(NfcEvent event)` method is called automatically when the `setNdefPushMessageCallback()` method is implemented and a mobile device is discovered to beam to.

On the other hand, in order to handle incoming beams, you need to implement three methods, which are `onResume()`, `onNewIntent(Intent intent)`, and `processIntent(Intent intent)`. The content of the first two methods can be the same in every activity that handles incoming beams, but the `onNewIntent()` method changes according to the activity since it will handle the incoming NDEF message.

AN ABSTRACT BEAM WITH SETNDEFPUSHMESSAGE( )

In the following code snippet, an abstract example to send and receive Android Beam using the `setNdefPushMessage()` method is given. At the beginning, an NDEF message is created and then the `setNdefPushMessage(message, this)` method is called. This way, when a mobile device is discovered to beam the data, the NDEF message that is sent to the `setNdefPushMessage()` method will be beamed to the target device. In order to handle incoming beams, the same methods are used with the `setNdefPushMessageCallback()` example: `onResume()`, `onNewIntent(Intent intent)`, and `processIntent(Intent intent)`.

```
public class PeertoPeerActivity extends Activity {
    @Override
    public void onCreate(Bundle savedInstanceState) {
        super.onCreate(savedInstanceState);
        setContentView(R.layout.main);
        mNfcAdapter = NfcAdapter.getDefaultAdapter(this);
        if (mNfcAdapter == null) {
            finish();
            return;
        }
        //You will write codes to create Ndef message here.
        mNfcAdapter.setNdefPushMessage(message, this);
    }

    @Override
    public void onResume() {
        super.onResume();
        if (NfcAdapter.ACTION_NDEF_DISCOVERED.equals(getIntent().getAction())) {
            processIntent(getIntent());
        }
    }

    @Override
    public void onNewIntent(Intent intent) {
        setIntent(intent);
```

```
        }

    void processIntent(Intent intent) {
        //You will write codes to handle Ndef Message here.
    }
}
```

DECLARING INTENT FILTERS

In order for the application running in the target mobile phone to handle the incoming beam, you need to use intent filters. Remember that intent filters could be declared by using either the tag intent dispatch system or the foreground dispatch system in reader/writer mode. However, you cannot use the foreground dispatch system in peer-to-peer mode. This means that you cannot declare any intent filter inside the activity but need to declare your intent filters in the manifest file instead. In other words, you need to use the tag intent dispatch system.

Please refer to Chapter 5 for information about the tag intent dispatch system and how intent filters are declared in the manifest file.

USING ANDROID APPLICATION RECORDS IN PEER-TO-PEER MODE

You have already learned what an Android Application Record (AAR) is, and how it can be used in NDEF records. In peer-to-peer mode, you can also use AAR in order to guarantee to run your application.

When a device receives a beam with an AAR in it, the AAR overrides the tag intent dispatch system; then, the corresponding activity defined in AAR runs. In order to create an AAR, you should add it to any place in your NdefMessage. However, you should not make it the first NDEF record of the NDEF message, unless the AAR is the only NDEF record in the message. This is because the first record of the NDEF message is used to determine the MIME type of the data in the tag. An example usage of the AAR is:

```
NdefMessage message = new NdefMessage (new NdefRecord[] {
                        record1,
                        NdefRecord.createApplicationRecord("com.nfclab.peertoepeer")
});
```

Please refer to Chapter 5 for a detailed description of AARs.

AN EXAMPLE BEAM APPLICATION USING SETNDEFPUSHMESSAGECALLBACK( )

In this example, an NDEF message is created from a TNF_WELL_KNOWN with RTD_URI and is beamed to another mobile using the setNdefPushMessageCallback() method. The complete code of the example is given in Listing 7-1.

LISTING 7-1: Activity of Peer to Peer Application with setNdefPushMessageCallback( ) (PeertoPeer1\src\com\nfclab\peertopeer1\PeertoPeer1.java)

```java
public class PeertoPeer1 extends Activity implements CreateNdefMessageCallback {
    NfcAdapter mNfcAdapter;
    private TextView messageText;
    private String payload = "";

    @Override
    public void onCreate(Bundle savedInstanceState) {
        super.onCreate(savedInstanceState);
        setContentView(R.layout.main);
        messageText = (TextView) this.findViewById(R.id.messageText);
        mNfcAdapter = NfcAdapter.getDefaultAdapter(this);
        if (mNfcAdapter == null) {
            messageText.setText("NFC apdater is not available");
            finish();
            return;
        }
        messageText.setText("Touch another mobile to beam 'nfclab.com'!!!");
        mNfcAdapter.setNdefPushMessageCallback(this, this);
    }

    @Override
    public NdefMessage createNdefMessage(NfcEvent event) {
        byte[] uriField = "nfclab.com".getBytes(Charset.forName("US-ASCII"));
        byte[] payload = new byte[uriField.length + 1];
        payload[0] = 0x01;
        System.arraycopy(uriField, 0, payload, 1, uriField.length);
        NdefRecord URIRecord  = new NdefRecord(NdefRecord.TNF_WELL_KNOWN,
                                               NdefRecord.RTD_URI,
                                               new byte[0], payload);
        NdefMessage message = new NdefMessage(new NdefRecord[] { URIRecord });
        return message;
    }

    @Override
    public void onResume() {
        super.onResume();
        if (NfcAdapter.ACTION_NDEF_DISCOVERED.equals(getIntent().getAction())) {
            processIntent(getIntent());
        }
    }

    @Override
    public void onNewIntent(Intent intent) {
        setIntent(intent);
    }

    void processIntent(Intent intent) {
        NdefMessage[] messages = getNdefMessages(getIntent());
        for(int i=0; i<messages.length; i++){
            for(int j=0; j<messages[0].getRecords().length; j++){
                NdefRecord record = messages[i].getRecords()[j];
                payload = new String(record.getPayload(),1,
```

```java
                            record.getPayload().length-1,
                            Charset.forName("UTF-8"));
            messageText.setText( payload );
        }
    }
    LinearLayout.LayoutParams params = new LinearLayout.LayoutParams(
    LinearLayout.LayoutParams.WRAP_CONTENT,
    LinearLayout.LayoutParams.WRAP_CONTENT);
    Button button = new Button(this);
    this.addContentView(button, params);
    messageText.setText("");
    button.setText("Open Link: "+ payload );

    button.setOnClickListener(new View.OnClickListener()
    {
        public void onClick(View view)
        {
            Intent data = new Intent();
            data.setAction(Intent.ACTION_VIEW);
            data.setData(Uri.parse("http://www."+payload));
            try {
                startActivity(data);
            } catch (ActivityNotFoundException e) {
                return;
            }
        }
    });
}

NdefMessage[] getNdefMessages(Intent intent) {
    NdefMessage[] msgs = null;

    if (NfcAdapter.ACTION_NDEF_DISCOVERED.equals(intent.getAction()))
    {
        Parcelable[] rawMsgs = intent.getParcelableArrayExtra(
                                        NfcAdapter.EXTRA_NDEF_MESSAGES);
        if (rawMsgs != null)
        {
            msgs = new NdefMessage[rawMsgs.length];
            for (int i=0; i < rawMsgs.length; i++) {
                msgs[i] = (NdefMessage) rawMsgs[i];
            }
        }else {
            byte[] empty = new byte[] {};
            NdefRecord record = new NdefRecord(NdefRecord.TNF_UNKNOWN,
                                        empty, empty, empty);
            NdefMessage msg = new NdefMessage(new NdefRecord[] { record });
            msgs = new NdefMessage[] { msg };
        }
    }else {
        Log.d("PeertoPeer1 ", "Unknown intent.");
        finish();
    }
    return msgs;
}
}
```

At the beginning, the activity implements `CreateNdefMessageCallback`. In the `onCreate()` method, the NFC adapter of the mobile is received by the `NfcAdapter` object and checked if the adapter is available. Then the `setNdefPushMessageCallback()` method is called. At this stage, the mobile waits to discover another mobile in order to beam to it (see Figure 7-2). When the mobile phone is touched to another mobile to beam, the `createNdefMessage()` method is invoked. Inside this method, an NDEF record of the `TNF_WELL_KNOWN` with `RTD_URI` type is created as shown in Chapter 5. Then an NDEF message is created using this record type. Returning the created NDEF message in the `createNdefMessage()` method enables the beaming of the message. At this point, the Touch to Beam UI is displayed on the screen (see Figure 7-3).

FIGURE 7-2

FIGURE 7-3

The `onResume()` and `onNewIntent()` methods are used in a similar way to the abstract examples given above. In the `processIntent()` method, first the `getNdefMessages()` method is invoked to extract the incoming NDEF message from the received intent. (The `getNdefMessages()` method is described and used in Chapter 5.) Then the payload is extracted from the NDEF message. The rest of the code is optional and can be personalized based on your requirements and design. In Listing 7-1, a new layout view and button are added to the layout (see Figure 7-4). When the button is clicked, the URL in the payload is displayed in a browser by creating an intent (see Figure 7-5).

FIGURE 7-4

FIGURE 7-5

In Listing 7-2, the manifest file of the example is given. You can see that NFC permission is added to the manifest file and an intent filter to filter RTD_URI data is also added to the activity.

LISTING 7-2: Manifest File of Peer-to-Peer Application with setNdefPushMessageCallback( )
(PeertoPeer1\AndroidManifest.xml)

```xml
<?xml version="1.0" encoding="utf-8"?>
<manifest xmlns:android="http://schemas.android.com/apk/res/android"
 package="com.nfclab.peertopeer1"
 android:versionCode="1"
 android:versionName="1.0" >
    <uses-sdk android:minSdkVersion="15" />
    <uses-permission android:name="android.permission.NFC"/>

    <application
     android:icon="@drawable/ic_launcher"
     android:label="@string/app_name" >
       <activity
        android:name=".PeertoPeer1"
        android:label="@string/app_name" >
         <intent-filter>
            <action android:name="android.intent.action.MAIN" />
            <category android:name="android.intent.category.LAUNCHER" />
         </intent-filter>
         <intent-filter>
            <action android:name="android.nfc.action.NDEF_DISCOVERED"/>
            <category android:name="android.intent.category.DEFAULT"/>
            <data android:scheme="http"
             android:host="*"
             android:pathPrefix="" />
         </intent-filter>
       </activity>
    </application>

</manifest>
```

AN EXAMPLE BEAM APPLICATION USING SETNDEFPUSHMESSAGE( )

In this example, an NDEF message is created from a TNF_WELL_KNOWN with RTD_TEXT type record and is beamed to another mobile using the setNdefPushMessage() method. The complete code of the example is given in Listing 7-3.

LISTING 7-3: Activity of Peer-to-Peer Application with setNdefPushMessage( ) (PeertoPeer2\src\
      com\nfclab\peertopeer2\PeertoPeer2.java)

```java
public class PeertoPeer2 extends Activity  {
    NfcAdapter mNfcAdapter;
    private TextView messageText;
    private String payload = "";
    private  EditText inputEditText;
    byte statusByte;

    @Override
    public void onCreate(Bundle savedInstanceState) {
        super.onCreate(savedInstanceState);
        setContentView(R.layout.main);
        messageText = (TextView) this.findViewById(R.id.messageText);
        mNfcAdapter = NfcAdapter.getDefaultAdapter(this);
        if (mNfcAdapter == null) {
            messageText.setText("NFC apdater  is not available");
            finish();
            return;
        }
        messageText.setText("Write some text to share");
    }

    public void onClickHandler(View view) {
        if(view.getId() == R.id.shareButton){
        inputEditText = (EditText)this.findViewById(R.id.inputEditText);
        String inputText = inputEditText.getText().toString();
        NdefMessage message=create_RTD_TEXT_NdefMessage(inputText);
        mNfcAdapter.setNdefPushMessage(message, this);
        Toast.makeText(this, "Touch another mobile to share the message",
                    Toast.LENGTH_SHORT).show();
        }
    }

    @Override
    public void onResume() {
        super.onResume();
        if (NfcAdapter.ACTION_NDEF_DISCOVERED.equals(getIntent().getAction())) {
            processIntent(getIntent());
        }
    }

    @Override
    public void onNewIntent(Intent intent) {
        setIntent(intent);
    }

    void processIntent(Intent intent) {
        NdefMessage[] messages = getNdefMessages(getIntent());
        for(int i=0; i<messages.length; i++)
        {
            for(int j=0; j<messages[0].getRecords().length; j++)
            {
```

```java
            NdefRecord record = messages[i].getRecords()[j];
            statusByte = record.getPayload()[0];
            int languageCodeLength= statusByte & 0x3F;
            int isUTF8 = statusByte-languageCodeLength;
            if(isUTF8 == 0x00)
            {
                payload = new String(
                                record.getPayload(),1+languageCodeLength,
                                record.getPayload().length-1-languageCodeLength,
                                Charset.forName("UTF-8"));
            }
            else if (isUTF8==-0x80)
            {
                payload = new String
                                (record.getPayload(),
                                 1+languageCodeLength,
                                 record.getPayload().length-1-languageCodeLength,
                                 Charset.forName("UTF-16")
                                );
            }
          messageText.setText("Text received: "+ payload);
        }
    }
}


NdefMessage create_RTD_TEXT_NdefMessage(String inputText)
{
    Locale locale = new Locale("en","US");
    byte[] langBytes = locale.getLanguage().getBytes(
                                        Charset.forName("US-ASCII"));
    boolean encodeInUtf8 = false;
    Charset utfEncoding = encodeInUtf8 ?
                        Charset.forName("UTF-8"):Charset.forName("UTF-16");
    int utfBit = encodeInUtf8 ? 0 : (1 << 7);
    byte status = (byte) (utfBit + langBytes.length);
    byte[] textBytes = inputText.getBytes(utfEncoding);
    byte[] data = new byte[1 + langBytes.length + textBytes.length];
    data[0] = (byte) status;
    System.arraycopy(langBytes, 0, data, 1, langBytes.length);
    System.arraycopy(textBytes, 0, data, 1 + langBytes.length, textBytes.length);
    NdefRecord textRecord = new NdefRecord(NdefRecord.TNF_WELL_KNOWN,
    NdefRecord.RTD_TEXT, new byte[0], data);
    NdefMessage message= new NdefMessage(new NdefRecord[] { textRecord});
    return message;
}

NdefMessage[] getNdefMessages(Intent intent)
{
    NdefMessage[] msgs = null;
    if (NfcAdapter.ACTION_NDEF_DISCOVERED.equals(intent.getAction()))
    {
        Parcelable[]
        rawMsgs = intent.getParcelableArrayExtra(NfcAdapter.EXTRA_NDEF_MESSAGES);
        if (rawMsgs != null)
```

continues

LISTING 7-3 *(continued)*

```
            {
                msgs = new NdefMessage[rawMsgs.length];
                for (int i=0; i < rawMsgs.length; i++) {
                    msgs[i] = (NdefMessage) rawMsgs[i];
                }
            } else {
                byte[] empty = new byte[] {};
                NdefRecord record = new NdefRecord(NdefRecord.TNF_UNKNOWN,
                                              empty, empty, empty);
                NdefMessage msg = new NdefMessage(new NdefRecord[] { record });
                msgs = new NdefMessage[] { msg };
            }
        } else {
            Log.d("Peer to Peer 2", "Unknown intent.");
            finish();
        }
        return msgs;
    }
}
```

As you can see in this code, the activity does not implement `CreateNdefMessageCallback`, since it will not use the callback. As a result, the `setNdefPushMessage()` method is used but it is not called in the `onCreate()` method since this method receives an NDEF message as a parameter. Instead, an `EditText` object named `inputEditText` is created to get an input from the user, as can be seen in the Listing 7-4 layout file (see also Figure 7-6).

LISTING 7-4: Layout File of Peer-to-Peer Application with setNdefPushMessageCallback( )
(PeertoPeer2\res\layout\main.xml)

```xml
<?xml version="1.0" encoding="utf-8"?>
<LinearLayout xmlns:android="http://schemas.android.com/apk/res/android"
 android:layout_width="fill_parent"
 android:layout_height="fill_parent"
 android:orientation="vertical" >

    <TextView
     android:id="@+id/messageText"
     android:layout_width="match_parent"
     android:layout_height="wrap_content" />

    <EditText
     adroid:id="@+id/inputEditText"
     android:inputType="text"
     android:layout_width="match_parent"
     android:layout_height="wrap_content"
     android:ems="10" >
     <requestFocus/>
    </EditText>

    <Button
     android:id="@+id/shareButton"
```

```
        android:layout_width="wrap_content"
        android:layout_height="wrap_content"
        android:text="@string/shareMessage"
        android:onClick="onClickHandler" />

    </LinearLayout>
```

When the user inputs the text and clicks the "Share Message" button, the `onClickHandler()` method is invoked. Inside this method, the inputted text is sent to the method `create_RTD_TEXT_NdefMessage()` in order to create an NDEF message using the input. Then, `setNdefPushMessage()` is called since there is an NDEF message that you may now share (see Figure 7-6). At this stage, when the mobile phone discovers another mobile to beam to, the application tries to beam the NDEF message and the Touch to Beam UI is displayed on the screen (see Figure 7-7).

FIGURE 7-6

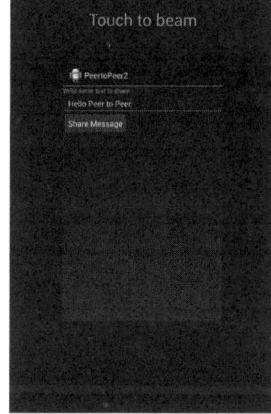

FIGURE 7-7

Remember that if your mobile phone discovers another mobile phone to beam to before you register the `setNdefPushMessage()` method, then it will beam a default NDEF message (the `URI` of your application).

The `create_RTD_TEXT_NdefMessage()` method creates an NDEF message based on the input text. (Creating `RTD_TEXT` is described in Chapter 5.)

The `onResume()` and `onNewIntent()` methods are used in the same fashion with the given abstract examples. As in the previous example, in the `processIntent()` method, first the `getNdefMessages()` method is called to extract the incoming NDEF message from the intent. (Please refer to Chapter 5 for detailed descriptions of the `getNdefMessages()` method.) After getting the NDEF message, the text inside the `RTD_TEXT` NDEF record is extracted. Afterwards, the text is displayed on the screen (see Figure 7-8). (Chapter 5 also describes how to extract the text from an `RTD_TEXT` NDEF record.)

FIGURE 7-8

Finally, in order to add the required intent filter, the following code snippet is added to the manifest file inside the `<activity>` … `</activity>` tags:

```
<intent-filter>
    <action android:name="android.nfc.action.NDEF_DISCOVERED" />
    <category android:name="android.intent.category.DEFAULT" />
    <data android:mimeType="text/plain" />
</intent-filter>
```

BEAM SUPPORT FOR API LEVEL 10

The `setNdefPushMessageCallback()` and `setNdefPushMessage()` methods are available in API level 14. In order to support older mobile devices that have an API level between 10 and 13, foreground NDEF Push is available, which provides similar functionality. Please notice that these APIs are deprecated and should not be used for devices with an API level of 14 and higher. The usage of this method is similar to the `setNdefPushMessage()` method.

Beaming with enableForegroundNdefPush( )

In order to beam using the `enableForegroundNdefPush()` method, follow these steps:

1. Create an NDEF message that will be beamed.

2. Call `enableForegroundNdefPush (Activity activity, NdefMessage message)`. When this method is called, it sets the received `NdefMessage` parameter as an NDEF message to beam. As the target mobile is discovered, the message is beamed to it.

3. Call `disableForegroundNdefPush(Activity)` in the `onPause()` method before the activity ends.

4. Call `enableForegroundNdefPush (Activity activity, NdefMessage message)` when the activity resumes in the `onPause()` method.

An Example Beam Application Using enableForegroundNdefPush( )

You will now modify the previous example that shares `TNF_WELL_KNOWN` with `RTD_TEXT` using `setNdefPushMessage()`. The complete code of the activity is given in Listing 7-5. The differences from the previous example, as highlighted in Listing 7-5, are:

1. Inside the `onClickHandler()` method, the `enableForegroundNdefPush()` method is called instead of the `setNdefPushMessage()` method.

2. The `onPause()` method is created and the `disableForegroundNdefPush()` method is called.

3. The content of the `onResume()` method is changed and the `enableForegroundNdefPush()` method is called.

LISTING 7-5: Activity of Peer-to-Peer Application with enableForegroundNdefPush( ) (PeertoPeer3\ src\com\nfclab\peertopeer3\PeertoPeer3.java)

```java
public class PeertoPeer3 extends Activity {
    NfcAdapter mNfcAdapter;
    private TextView messageText;
    private String payload = "";

    private EditText inputEditText;
    byte statusByte;
    private String inputText;
    private NdefMessage message;

    @Override
    public void onCreate(Bundle savedInstanceState) {
        super.onCreate(savedInstanceState);
        setContentView(R.layout.main);
        messageText = (TextView) this.findViewById(R.id.messageText);
        mNfcAdapter = NfcAdapter.getDefaultAdapter(this);
        if (mNfcAdapter == null) {
            messageText.setText("NFC apdater is not available");
            finish();
            return;
        }
        messageText.setText("Write some text to share");
    }

    public void onClickHandler(View view) {
        if( view.getId() == R.id.shareButton )
        {
            inputEditText = (EditText)this.findViewById(R.id.inputEditText);
            inputText = inputEditText.getText().toString();
            message = create_RTD_TEXT_NdefMessage(inputText);
            mNfcAdapter.enableForegroundNdefPush(this, message);
            Toast.makeText(this, "Touch another mobile to share the message",
                        Toast.LENGTH_SHORT).show();
        }
    }


    @Override
    public void onResume() {
        super.onResume();
        if (NfcAdapter.ACTION_NDEF_DISCOVERED.equals(getIntent().getAction())) {
            processIntent(getIntent());
        }
        else{
            mNfcAdapter.enableForegroundNdefPush(this,
                create_RTD_TEXT_NdefMessage(""));
        }
    }

    @Override
    public void onPause() {
```

continues

LISTING 7-5 *(continued)*

```java
        super.onPause();
        mNfcAdapter.disableForegroundNdefPush(this);
    }

    @Override
    public void onNewIntent(Intent intent) {
        setIntent(intent);
    }

    void processIntent(Intent intent) {
        NdefMessage[] messages = getNdefMessages(getIntent());
        for(int i=0; i<messages.length; i++){
        for(int j=0; j<messages[0].getRecords().length; j++)
        {
            NdefRecord record = messages[i].getRecords()[j];
            statusByte = record.getPayload()[0];
            int languageCodeLength= statusByte & 0x3F;
            int isUTF8 = statusByte-languageCodeLength;
            if( isUTF8 == 0x00 )
            {
                payload = new String( record.getPayload(),
                                1+languageCodeLength,
                                record.getPayload().length-1-languageCodeLength,
                                Charset.forName("UTF-8"));
            }
            else if ( isUTF8 == -0x80)
            {
                payload = new String( record.getPayload(),
                                1+languageCodeLength,
                                record.getPayload().length-1-languageCodeLength,
                                Charset.forName("UTF-16"));
            }
            messageText.setText("Text received: " + payload);
        }
    }

    NdefMessage create_RTD_TEXT_NdefMessage(String inputText)
    {
        Locale locale = new Locale("en","US");
        byte[] langBytes = locale.getLanguage().getBytes(
                                        Charset.forName("US-ASCII"));
        boolean encodeInUtf8 = false;
        Charset utfEncoding = encodeInUtf8 ?
                        Charset.forName("UTF-8") : Charset.forName("UTF-16");
        int utfBit = encodeInUtf8 ? 0 : (1 << 7);
        byte status = (byte) (utfBit + langBytes.length);
        byte[] textBytes = inputText.getBytes(utfEncoding);
        byte[] data = new byte[1 + langBytes.length + textBytes.length];
        data[0] = (byte) status;
        System.arraycopy(langBytes, 0, data, 1, langBytes.length);
```

```
                System.arraycopy(textBytes, 0, data, 1 + langBytes.length, textBytes.length);
                NdefRecord textRecord = new NdefRecord(NdefRecord.TNF_WELL_KNOWN,
                                                    NdefRecord.RTD_TEXT,
                                                    new byte[0], data);
                NdefMessage message = new NdefMessage(new NdefRecord[] { textRecord});
                return message;
        }

        NdefMessage[] getNdefMessages(Intent intent)
        {
            NdefMessage[] msgs = null;
            if (NfcAdapter.ACTION_NDEF_DISCOVERED.equals(intent.getAction()))
            {
                Parcelable[] rawMsgs = intent.getParcelableArrayExtra(
                                            NfcAdapter.EXTRA_NDEF_MESSAGES);
                if (rawMsgs != null) {
                    msgs = new NdefMessage[rawMsgs.length];
                    for (int i = 0; i < rawMsgs.length; i++)
                    {
                        msgs[i] = (NdefMessage) rawMsgs[i];
                    }
                }
                else
                {
                    byte[] empty = new byte[] {};
                    NdefRecord record = new NdefRecord( NdefRecord.TNF_UNKNOWN,
                                                    empty, empty, empty );
                    NdefMessage msg = new NdefMessage(new NdefRecord[] { record });
                    msgs = new NdefMessage[] { msg };
                }
            }
            else {
                Log.d("Peer to Peer 3", "Unknown intent.");
                finish();
            }
            return msgs;
        }
    }
}
```

ANDROID OS TO HANDLE THE INCOMING BEAM

When you create a well-known record, you can leave Android OS to do the work of receiving the beam. This time, Android OS will handle the incoming beam with its default OS programs.

The example given in Listing 7-6 demonstrates the Android OS's handling job process. In the example, a contact is selected from the phone's existing contact list, and beamed to another mobile. Since Android OS will handle a received contact with its existing contact application, this beam-receiving process is not implemented in the activity and is left to the receiving mobile phone's Android OS.

LISTING 7-6: Activity of Peer-to-Peer Contact Sharing (PeertoPeer4\src\com\nfclab\peertopeer4\
          PeertoPeer4.java)

```java
public class PeertoPeer4 extends Activity
{
    NfcAdapter mNfcAdapter;
    private TextView messageText;
    public Button sendButton;
    private final int PICK_CONTACT = 1;
    private byte[] bytesToSend;

    @Override
    public void onCreate(Bundle savedInstanceState)
    {
        super.onCreate(savedInstanceState);
        setContentView(R.layout.main);

        messageText = (TextView)findViewById(R.id.messageText);
        messageText.setText("Select a contact to share");

        mNfcAdapter = NfcAdapter.getDefaultAdapter(this);

        if (mNfcAdapter == null) {
            finish();
            return;
        }

        sendButton = (Button) findViewById(R.id.selectButton);
        sendButton.setOnClickListener(new OnClickListener()
        {
            public void onClick(View v)
            {
                Intent intent = new Intent(Intent.ACTION_PICK);
                intent.setType(ContactsContract.Contacts.CONTENT_TYPE);
                startActivityForResult(intent, PICK_CONTACT);
            }
        });
    }


    public void onActivityResult(int reqCode, int resultCode, Intent data)
    {
        super.onActivityResult(reqCode, resultCode, data);
        String contactInfo = "";
        switch (reqCode)
        {
            case (PICK_CONTACT):
            {
                if (resultCode == Activity.RESULT_OK)
                {
                    Uri contactData = data.getData();
                    Cursor people = getContentResolver().query(contactData,
                                                    null, null, null, null);
                    if ( people.moveToFirst() )
```

```
            {
              try
              {
                 String lookupKey = people.getString(people.getColumnIndex(
                                                  Contacts.LOOKUP_KEY));
                 Uri uri = Uri.withAppendedPath(
                     ContactsContract.Contacts.CONTENT_VCARD_URI, lookupKey);
                 AssetFileDescriptor afd;
                 try {
                   afd = getContentResolver().openAssetFileDescriptor
                                                  (uri, "r");
                   FileInputStream fis = afd.createInputStream();
                   bytesToSend = new byte[(int) afd.getDeclaredLength()];
                   fis.read(bytesToSend);
                   contactInfo = new String(bytesToSend);
                 } catch (FileNotFoundException e) {
                   e.printStackTrace();
                 } catch (IOException e) {
                   e.printStackTrace();
                 }
               } finally {
                 people.close();
               }

               NdefMessage message = create_MIME_NdefMessage(
                                          "text/x-vcard", bytesToSend);
               mNfcAdapter.setNdefPushMessage(message, this);
               messageText.setText(contactInfo);
               Toast.makeText(this, "Touch another mobile to share the contact",
               Toast.LENGTH_SHORT).show();
              }
           } if (resultCode == Activity.RESULT_OK)
        }
      }
    }

  public NdefMessage create_MIME_NdefMessage(String mimeType, byte[] payload)
  {
     NdefRecord mimeRecord = new NdefRecord(NdefRecord.TNF_MIME_MEDIA,
                               mimeType.getBytes(), new byte[0], payload);
     NdefMessage message = new NdefMessage(new NdefRecord[] { mimeRecord });
     return message;
  }

  @Override
  public void onResume() {
     super.onResume();
     if (!mNfcAdapter.isEnabled())
     {
        startActivity(new Intent(Settings.ACTION_NFC_SETTINGS));
     } else if (!mNfcAdapter.isNdefPushEnabled()) {
        startActivity(new Intent(Settings.ACTION_NFCSHARING_SETTINGS));
     }
  }
}
```

In the opening screen (see Figure 7-9), a `TextView` and a button are created. When the user clicks the button, the contact list is displayed and the application waits for the user to select a contact to send (see Figure 7-10). When the user selects a contact, the contact information is extracted from the database as a file input stream and saved into a byte array named `bytesToSend`. This byte array is then sent to the `create_MIME_NdefMessage()` method in order to create the NDEF message to beam. Since the contact has a `vCard` file format, the MIME type of the record is set as `text/x-vcard`. When the NDEF message is created, the `setNdefPushMessage()` method is called to set the NDEF message as a beam. Then, the mobile waits for the user to touch another mobile to beam the message (see Figure 7-11).

FIGURE 7-9

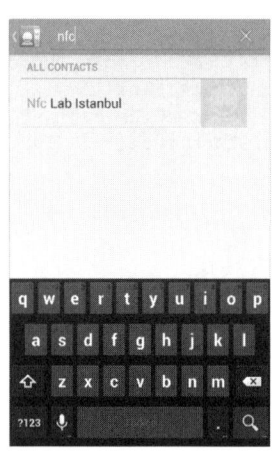

FIGURE 7-10

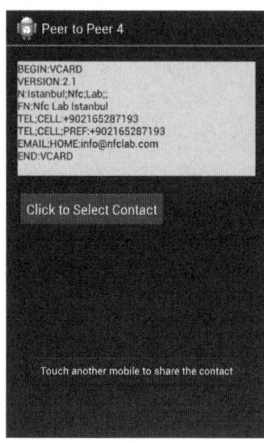

FIGURE 7-11

After the message is beamed, the receiving phone's Android OS handles the received beam since the received data is a known type (`text/x-vcard`). Then it adds the received contact to its contact list as shown in Figure 7-12. Moreover, you must not define any intent filter, since defining the corresponding intent filter will get the incoming beam to the activity, and Android OS will not be able to process the contact by itself.

Additionally, in the `onResume()` method, the NFC and Android Beam settings are checked. The activity first checks if the NFC is turned on or off from the Android settings. If it is turned off, the activity runs the Wireless & Networks settings menu to force the user to enable NFC (see Figure 7-13). After the user enables NFC, it checks if Android Beam is turned on or off. If Android Beam is turned off, the activity runs Android Beam settings to force the user to enable it (see Figure 7-14). Note that the `isNdefPushEnabled()` method and `ACTION_NFC_SETTINGS` constant require an API level of 16.

FIGURE 7-12

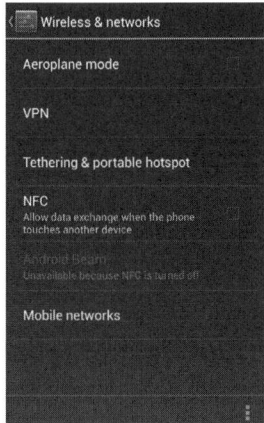

FIGURE 7-13

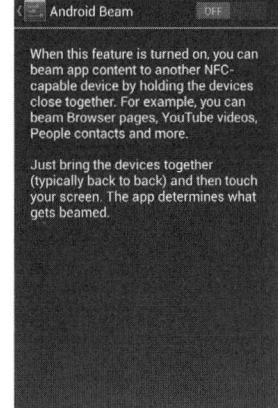

FIGURE 7-14

BEAMING FILES

In order to beam files such as images, videos, and so on, new methods for sending beams are added in API level 16. In order to beam files, you first need to save the file's URI to be beamed into a URI array, and then use the required methods described in this section. When the mobiles touch each other, the Bluetooth of the mobiles are activated and the file transfer is performed over Bluetooth.

In order to beam URIs, you need to provide the file's `file` scheme or `content` scheme. As with beaming NDEF messages, there are two options to beam URIs: you can implement either with a callback (`setBeamPushUrisCallback()` method) or without a callback (the `setBeamPushUris()` method). `setBeamPushUrisCallback()` has priority over `setBeamPushUris()`. So if both methods are implemented in the activity, `setBeamPushUrisCallback()` will take the priority and the activity will use the callback.

When both beaming an NDEF message and beaming a URI are implemented, `setBeamPushUris()` and `setBeamPushUrisCallback()` both have the priority over `setNdefPushMessage()` and `setNdefPushMessageCallback()`.

Beaming with setBeamPushUrisCallback( )

In order to beam using the `setBeamPushUrisCallback()` method, you should follow these steps:

1. Implement `CreateBeamUrisCallback` in your activity.

2. Call `setBeamPushUrisCallback(NfcAdapter.CreateBeamUrisCallback callback, Activity activity)`. When this method is called, the activity accepts a callback and if a mobile device is discovered to beam the data, the `createBeamUris(NfcEvent)` method is invoked automatically.

3. Inside the `createBeamUris(NfcEvent)` method, create the URI array and return it. The files in the array will be beamed to the target device.

When the target mobile phone is discovered, the `createBeamUris(NfcEvent)` method is invoked, which returns a URI array. When two mobiles touch, then the Touch to Beam UI is displayed on top of your activity. At this time, when the user touches the screen, the Bluetooth connection is established between two mobiles and the file is sent to the target mobile.

Beaming with setBeamPushUris( )

In order to beam using the `setBeamPushUris()` method, you should follow these steps:

1. Create a URI array.

2. Call `setBeamPushUris(Uri[] uris, Activity activity)`. When this method is called, the activity prepares the files to be sent from the received URI array. When the target mobile is discovered, it beams the file.

The `setBeamPushUris()` method normally sets the URI array statically because it gets the array as a parameter. You may sometimes prefer to use it instead of `setBeamPushUrisCallback()` based on your implementation.

An Example Beam Application Using setBeamPushUrisCallback( )

In this example, a URI of an image file in the phone's storage is gathered and sent to a target mobile phone using the `setBeamPushUrisCallback()` method. First the activity implements `CreateBeamUrisCallback`:

```
public class UriShareActivity extends Activity implements CreateBeamUrisCallback {
```

Then the `setBeamPushUrisCallback` method is implemented inside the `oncreate()` method:

```
mNfcAdapter.setBeamPushUrisCallback(this, this);
```

At this stage, the mobile waits to discover the target mobile to beam. When it discovers the target mobile, the activity calls `createBeamUris()` method and beams the returning URI array:

```
@Override
public Uri[] createBeamUris(NfcEvent event) {
    Uri uri = Uri.parse("file:///sdcard/Pictures/Screenshots/1.png");
    Uri[] uriArray = {uri};
    return uriArray;
}
```

When the `uriArray` is beamed to the target mobile, the Bluetooth connection is activated and the file is sent over Bluetooth. The `onResume()` and `onNewIntent()` methods are used in a similar way to the previous peer-to-peer applications.

An Example Beam Application using setBeamPushUris( )

This time, the activity does not implement `CreateBeamUrisCallback`, since it will use the `setBeamPushUris()` method. You need to implement the `setBeamPushUris()` as follows whenever you want to beam the data:

```
Uri uri = Uri.parse("file:///sdcard/Pictures/Screenshots/1.png");
Uri[] uriArray = {uri};
mNfcAdapter.setBeamPushUris(uriArray, this);
```

At this stage, when the mobile discovers the target, the Touch to Beam UI is displayed on the screen. When the user touches the screen, the data is beamed to the target mobile.

The onResume() and onNewIntent() methods are used in the same fashion with the above examples. As in the previous example, when the uriArray is beamed to the target mobile, the Bluetooth connection is activated and the file is sent over Bluetooth.

SUMMARY

Peer-to-peer mode allows two mobiles to exchange data when they touch each other. Sending data using peer-to-peer mode in Android is named Android Beam. In order to beam, there are some preconditions. First of all, Android Beam must be turned on from the settings in both mobile phones. Then the activity in the first mobile phone must be in the foreground and the second mobile phone's screen must not be locked. When these conditions are met, the first mobile can beam to the second mobile.

In exactly the same way as it is done in reader/writer mode, you should create NDEF messages to beam to the other device. In reader/writer mode, you create an NDEF message and then write to a tag or read an NDEF message from a tag. In peer-to-peer mode, you create an NDEF message and then beam it to another mobile or read an incoming beam; in other words, you read an incoming NDEF message. So, in order to beam, you need to encapsulate whatever data you want to an NdefMessage object.

When your activity implements an Android Beam, the tag intent dispatch system is disabled so that you cannot scan any tag. However, if you also implement foreground dispatching in your activity, your activity will also be able to scan tags.

There are two different ways to implement Android Beam. The first is to implement the setNdefPushMessageCallback() method; the second is to implement the setNdefPushMessage() method. The setNdefPushMessageCallback() method dynamically generates NDEF messages, whereas the setNdefPushMessage() method sets it statically. You may call both of the methods anywhere in your activity. setNdefPushMessageCallback() has priority over setNdefPushMessage() so, if both are implemented in the activity, setNdefPushMessageCallback() will take priority.

Both methods are available at API level 14. In order to support mobile devices that have an API level between 10 and 14, foreground NDEF Push is available, which provides similar functionality. However, foreground NDEF Push is deprecated in API level 14 and should not be used when API level 14 and higher is available.

Peer-to-Peer Mode Applications

WHAT'S IN THIS CHAPTER?

- ➤ Use cases for peer-to-peer mode applications
- ➤ NFC chatting
- ➤ NFC Guess Number
- ➤ NFC Panic Bomb

WROX.COM CODE DOWNLOADS FOR THIS CHAPTER

The wrox.com code downloads for this chapter are found at www.wrox.com/remtitle
.cgi?isbn=1118380096 on the Download Code tab. The code is in the Chapter 8 download
and individually named according to the names throughout the chapter.

In this chapter, three peer-to-peer use cases are implemented using Android APIs: chatting
between two parties, a number guessing game, and a ticking time bomb game.

The first use case is a chatting application in which users can write chat messages and send
them to others by touching their mobile phones. This use case can be thought of as a chat
application in which two people exchange information in a location where many people exist
but where nobody else can hear their communication.

The second use case is a game about guessing a secret number. The application randomly
generates an integer between 1 and 100, and two or more players in the group try to guess
the number. After the users have made their guesses, the application directs them to suggest
a higher or lower number next time. The game ends when one of the players has guessed the
secret number.

The third use case is a game in which a bomb explodes after a randomly selected time by the application. The user who is holding the bomb can pass it to another player only after answering a question asked by the mobile. The user who is holding the bomb when it explodes loses the game.

NFC CHATTING

NFC chatting is an interesting use case of peer-to-peer mode, enabling one-to-one instant messaging between two users. One user writes a message, then touches a friend's mobile to send the message to them. The application technically prepares an NDEF message, consisting of the chat message, and then beams the NDEF message to the target device. When the target device receives the beam, it displays the chat message on the screen. The users may continue exchanging chat messages as often as they wish, and in any order.

There are two Java classes in the implemented project: ChatAdapter and ChatActivity. The purpose of ChatAdapter is to maintain a list of the exchanged messages. It internally uses an ArrayList of String items, or messages. When a message is sent or received, it is added to the list of messages by invoking the void add(String message) method. Existing messages in the database can be questioned using the String getItem(int index) method. int getCount() returns the number of existing messages in the list, and View getView() is used to properly display the messages on the screen. Incoming messages are aligned to the left, and outgoing messages are aligned to the right. Incoming messages are enclosed from both left and right using < tokens, and outgoing messages are enclosed using > tokens in order to distinguish them. When displaying the messages, only the first token that is embedded at the beginning of a message is displayed, while the token at the end of a message is kept hidden. The complete code of the ChatAdapter class is given in Listing 8-1.

LISTING 8-1: ChatAdapter Class (NFCChat\src\com\nfclab\chat\ChatAdapter.java)

```java
public class ChatAdapter extends ArrayAdapter<String> {

    private TextView messageTV;
    private List<String> messagesAL = new ArrayList<String>();
    private LinearLayout lineLL;
    int size;

    @Override
    public void add(String message) {
       messagesAL.add(message);
    }

    public ChatAdapter(Context context, int textViewResourceId, int size) {
       super(context, textViewResourceId);
       this.size = size;
    }

    public int getCount() {
```

```
            return this.messagesAL.size();
        }

        public String getItem(int index) {
            return this.messagesAL.get(index);
        }

        public View getView(int position, View convertView, ViewGroup parent)
        {
            View row = convertView;
            if (row == null) {
                LayoutInflater inflater = (LayoutInflater)
                    this.getContext().getSystemService(Context.LAYOUT_INFLATER_SERVICE);
                row = inflater.inflate(R.layout.message, parent, false);
            }

            lineLL = (LinearLayout) row.findViewById(R.id.lineLL);

            String oneMessage = getItem(position);

            messageTV = (TextView) row.findViewById(R.id.messageTV);
            // oneMessage.length()-1 hides the last token in the chat message
            messageTV.setText(oneMessage.substring(0,oneMessage.length()-1));
            messageTV.setBackgroundResource(R.drawable.bubble_banana);

            if ( oneMessage.charAt(0) == '<')
                lineLL.setGravity(Gravity.LEFT);
            else
                lineLL.setGravity(Gravity.RIGHT);
            return row;
        }
    }
```

The second class is ChatActivity. The activity mainly maintains a user interface to enable sending a message, and a ListView to display the sent and received messages on the screen. It keeps a list of the exchanged messages using the ChatAdapter class, persistently stores the same list of messages using SharedPreferences, and uses a string variable, namely messageList, to handle the same list of messages internally.

As a design criterion, if the user quits the application and opens it again, previous messages are to be displayed on the screen, which requires saving the exchanged messages in a persistent storage. SharedPreferences is used for persistent storage of the messages. In order to satisfy that criterion, all messages are saved in a persistent database using the SharedPreferences.Editor interface via the saveMessages() method. The following code from the ChatActivity class saves the messages:

```
String PREFERENCE_NAME = "NFCChat";
SharedPreferences.Editor editor;
SharedPreferences settings;

@Override
public void onCreate(Bundle savedInstanceState)
{
    settings = getSharedPreferences(PREFERENCE_NAME, MODE_PRIVATE);
```

```
    messageList = settings.getString("messageList", "");

    String thisMessage;
    char token;
    int start = 0;
    int end;
    while ( messageList.length() > 0 ) {

        //get the token, which is either '<' or '>'
        token = messageList.substring(start, start+1).charAt(0);

        //find the index of the token at the end of thisMessage
        end = messageList.substring(start+1).indexOf(token) + 1;

        //retrieve thisMessage from messageList
        thisMessage = messageList.substring(start, end+1);

        //add thisMessage to the list of messages that is maintained by adapter
        adapter.add(thisMessage);

        //start of next message is the next index after end of current message.
        start = end + 1;
        if ( messageList.substring(start) != null )

            //if there more unprocessed messages, clip the remaining messages
            messageList = messageList.substring(start);
        else

            //there are no more messages
            messageList = "";
        start = 0; //the story continues
    }
}


protected void saveMessages() {
    messageList = "";

    //append all messages in the adapter to messageList
    for ( int i = 0; i < adapter.getCount(); i++)
      messageList += adapter.getItem(i);

    //prepare editor to save the messageList as a string, and save it.
    editor = settings.edit();
    editor.putString("messageList", messageList);
    editor.commit();
}

public void get(String messageToGet)
{
    //append all messages in the adapter to messageList
    for ( int i = 0; i < adapter.getCount(); i++)
        messageList += adapter.getItem(i);

    //add the new message to the adapter
```

```
        //notice that '<' is appended to the start and end of the message,
        //the first token denotes that the message is a received message
        //and will be left aligned. The second token marks the end of the message
        //and is not displayed on the screen.
        adapter.add('<' + messageToGet + '<');

        //re-generate messageList by using the messages in the adapter,
        // including the most recent one.
        messageList = "";

        for ( int i = 0; i < adapter.getCount(); i++)
            messageList += adapter.getItem(i);

        //all messages are saved to the adapter after each new sent or received message
        saveMessages();
    }

    public boolean onKey(View v, int keyCode, KeyEvent event)
    {
        if ((event.getAction() == KeyEvent.ACTION_DOWN)
            && (keyCode == KeyEvent.KEYCODE_ENTER)) //message is inputted
        {
            if( messageToSendET.getText().length() > 0 )
            {
                messageToSend = messageToSendET.getText().toString();
                adapter.add('>' + messageToSend + '>');
                saveMessages();
                messageToSendET.setText("");
                Toast.makeText(this,
                            "Touch another mobile to share the chat message",
                            Toast.LENGTH_LONG).show();
            }
            return true;
        }
        return false;
    }
}
```

FIGURE 8-1

When the application is launched, the main screen contains ListView and EditText objects (see Figure 8-1). The ListView object is used to display the sent and received chat messages, and the EditText object is used to get the chat message input from a user.

When a user types a chat message and presses the Enter button from the keyboard, the text is saved in a messageToSend string that will be beamed when the mobiles touch each other (see Figure 8-2). After the chat message is beamed to the target device (see Figure 8-3), the chat application on the target device is invoked and displays the chat message on the screen (see Figure 8-4). The process continues in this way until the parties wish to quit (see Figure 8-5). See Figure 8-6 for the activity flow diagram of the application.

FIGURE 8-2

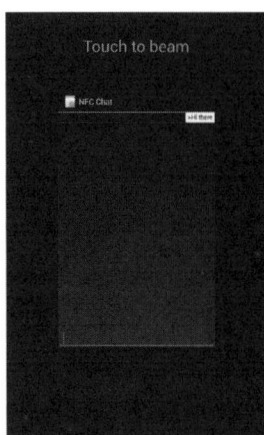

FIGURE 8-3

FIGURE 8-4

FIGURE 8-5

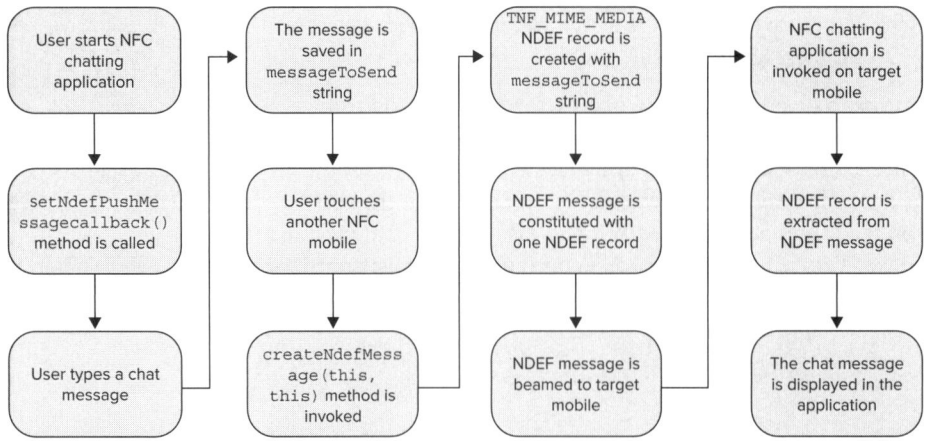

FIGURE 8-6

The NFC part of the application is developed with `NdefMessageCallback` so that activity implements `CreateNdefMessageCallback`:

```
public class ChatActivity extends Activity
    implements CreateNdefMessageCallback, OnKeyListener {
```

In the `onCreate()` method, the `setNdefPushMessageCallback()` method is implemented using the following code:

```
mNfcAdapter.setNdefPushMessageCallback(this, this);
```

Since the NFC part is implemented with the callback, you need to define the `createNdefMessage()` method that will be invoked when the mobile device starts to beam. In this method, you need to create the NDEF message and return it in order to beam the message. `TNF_MIME_MEDIA` TNF is used to format NDEF records, so in order to create the NDEF message, another method is defined and named `create_MIME_NdefMessage()`. The required data are sent to this method, which are the MIME media type and the chat message. This method first creates the NDEF record and then the NDEF message according to the `TNF_MIME_MEDIA` rules and sends the NDEF message back to the `createNdefMessage()` method. Then the NDEF message is returned in the `createNdefMessage()` method, which will automatically beam it when a target is discovered. Furthermore, `messageToSend` is set to an empty string, since the already sent messages do not need to be sent again:

```
@Override
public NdefMessage createNdefMessage(NfcEvent event)
{
    NdefMessage message = create_MIME_NdefMessage("application/nfcchat",
                                            messageToSend);
    messageToSend = "";
    return message;
}


public NdefMessage create_MIME_NdefMessage(String mimeType, String payload)
{
    NdefRecord mimeRecord = new NdefRecord(NdefRecord.TNF_MIME_MEDIA,
                            mimeType.getBytes(), new byte[0],
                            payload.getBytes(
                                    Charset.forName("US-ASCII"))
                            );
    NdefMessage message = new NdefMessage(new NdefRecord[] { mimeRecord});
    return message;
}
```

On the receiving side, you have already learned that you need to define three methods: `onResume()`, `onNewIntent(Intent intent)`, and `processIntent(Intent intent)`. The `onResume()` and `onNewIntent()` methods handle the incoming beams, and the `processIntent()` method handles the incoming NDEF message by calling the `getNdefMessages()` method:

```
@Override
public void onResume()
{
    super.onResume();
    if (NfcAdapter.ACTION_NDEF_DISCOVERED.equals(getIntent().getAction()))
    {
```

```
        processIntent(getIntent());
    }
}

@Override
public void onNewIntent(Intent intent) {
    setIntent(intent);
}


void processIntent(Intent intent)
{
    NdefMessage[] messages = getNdefMessages(getIntent());
    for(int i=0; i<messages.length; i++)
    {
        for(int j=0; j<messages[0].getRecords().length; j++)
        {
            NdefRecord record = messages[i].getRecords()[j];
            String payload = new String(record.getPayload());
            get( payload );
        }
    }
}
```

In addition, the get() method is called in the processIntent() method, which puts the chat message on the screen in a desired display setting. You can personalize the get() method of the code, since it affects the screen orientation and can be implemented in many other ways. The code of the get() method is shown in the preceding code snippet.

Inside the manifest file, you need to add an NFC permission in order to use the NFC adapter. Furthermore, you need to define the required intent filter to run the application that receives the beam. Since the application uses a TNF_MIME_MEDIA TNF for NDEF records, you need to define the corresponding intent filter. The manifest file of the application is given in Listing 8-2.

> **NOTE** *Please see Chapter 5, "NFC Programming: Reader/Writer Mode," for detailed information on* TNF_MIME_MEDIA.

LISTING 8-2: Manifest File of the NFC Chat Project (NFCChat\AndroidManifest.xml)

```xml
<?xml version="1.0" encoding="utf-8"?>
<manifest xmlns:android="http://schemas.android.com/apk/res/android"
 package="com.nfclab.chat"
 android:versionCode="1"
 android:versionName="1.0" >

    <uses-sdk android:minSdkVersion="15" />
    <uses-permission android:name="android.permission.NFC" />
    <application
     android:icon="@drawable/ic_launcher"
     android:label="@string/app_name" >
        <activity
```

```
            android:name=".ChatActivity"
            android:label="@string/app_name" >
              <intent-filter>
                 <action android:name="android.intent.action.MAIN" />
                 <category android:name="android.intent.category.LAUNCHER" />
              </intent-filter>
              <intent-filter>
                 <action android:name="android.nfc.action.NDEF_DISCOVERED" />
                 <category android:name="android.intent.category.DEFAULT" />
                 <data android:mimeType="application/nfcchat" />
              </intent-filter>
           </activity>
       </application>
   </manifest>
```

NFC GUESS NUMBER

The NFC Guess Number use case is an entertainment game in which two or more players try to find a secret number. The secret number is initially generated by the mobile phone in the interval of 1–100 using a random number generator. A player makes a guess at the secret number. If the guess is incorrect, the application directs the player to make a higher or lower guess. Then the player touches their mobile to the next player, and the secret number and previous guesses are beamed to the target mobile. If the new player makes an incorrect guess, that player beams back to the first player, and so on. The game ends when one of the players finds the secret number.

When the application is initially launched, the main screen contains EditText objects, TextView objects, and a Button object, as shown in Figure 8-7. This main screen displays the previous guesses, the possible high and low guesses that can be made, and a button to submit the guess. The player can also input their guess using components on this screen. When the game is initially launched on any player's mobile, the secret number between 1 and 100 is randomly generated. When the player inputs their guess (see Figure 8-8) and submits it by clicking the "Guess" button, the application checks the guess and states whether it is low or high (see Figure 8-9). At the same time, high and low number fields are updated for the player who will be making the second guess.

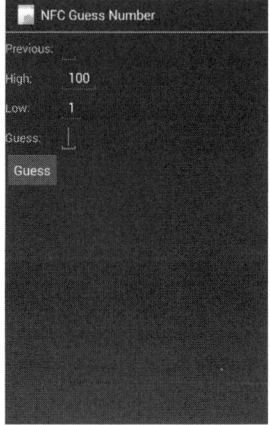

FIGURE 8-7

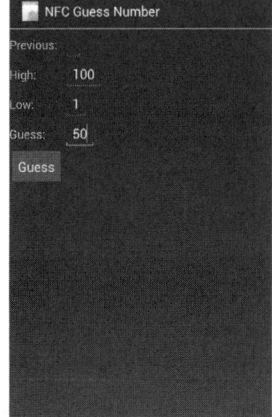

FIGURE 8-8

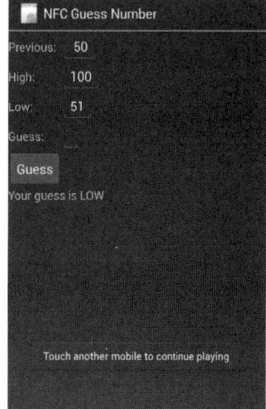

FIGURE 8-9

After a player makes an incorrect guess, the application asks that player to touch their mobile to the next player's mobile. At this point, the incorrect guesses previously made by all players, the low and high bounds of the range for the next guess, the secret number, and the secret number's find status are packaged into an NDEF message and beamed to the target mobile. When the second player receives the beam, the application is launched automatically. The previous guesses and the range for the next guess are displayed by the application using the received beam (see Figure 8-10). Then the application waits for the user to guess the number. The application continues in this way until the secret number is successfully found (see Figure 8-11), at which point the application displays a success message (see Figure 8-12). If the mobile phone beams after this time, the target mobile displays a message that a user has already found the secret number, by using the secret number's find status in the beam (see Figure 8-13). See Figure 8-14 for the activity flow diagram of the application.

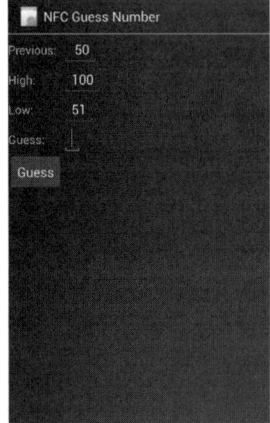

FIGURE 8-10

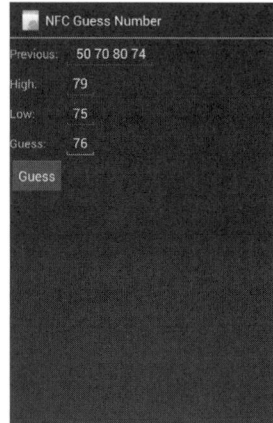

FIGURE 8-11

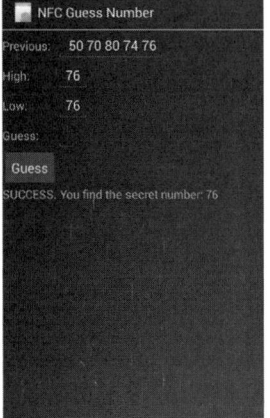

FIGURE 8-12

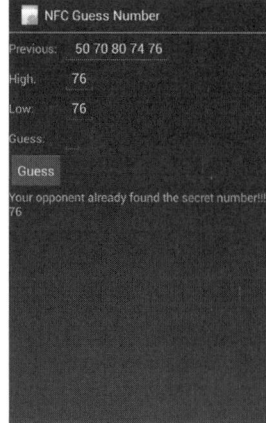

FIGURE 8-13

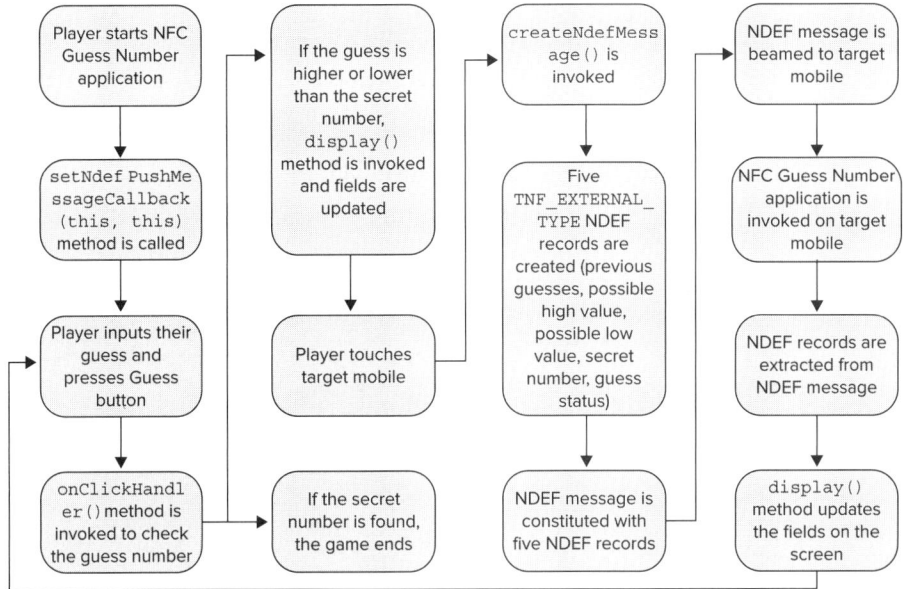

FIGURE 8-14

The NFC part of the application is developed with NdefMessageCallback so that activity implements CreateNdefMessageCallback:

```
public class GuessNumberActivity extends Activity
                             implements CreateNdefMessageCallback {
```

In the onCreate() method, the setNdefPushMessageCallback() method is implemented using the following code:

```
mNfcAdapter.setNdefPushMessageCallback(this, this);
```

Then, the target integer that stores the secret number is randomly generated between 1 and 100 using the random() method. The display() method is then called, which updates the screen objects:

```
if ( target == 0 ) {
   target = random();
}
display(true);

public int random() {
   return (new Random().nextInt(100)) + 1;
}

public void display(boolean focus)
{
   previousET.setText(previous);
   highET.setText("" + high);
   lowET.setText("" + low);
   guessET.setText("");
   guessET.setFocusable(focus);
}
```

When the player inputs a guess and clicks the "Guess" button, the `onClickHandler()` method is invoked, which checks the guess. There are three options: the secret number is found, the guess is higher than the secret number, or the guess is lower than the secret number. When the secret number is found, the application displays a success message and changes the found boolean value to true. If the guess is wrong, the low or high values are updated appropriately, and a message to direct the user is displayed on the screen. Finally, the `display()` method is called, which updates the fields on the screen and displays a `Toast` to inform the player to touch the mobile to the next player's mobile:

```
public void onClickHandler(View view)
{
    if(view.getId() == R.id.guessButton)
    {
        guess = Integer.parseInt(guessET.getText().toString());
        previous = previous + " " + guessET.getText();
        if ( guess == target )
        {
            low = guess;
            high = guess;
            found = true;
            resultTV.setText("SUCCESS. You find the secret number: "+guess);
        }
        else
        {
            if ( guess < target )
            {
                low = guess + 1;
                resultTV.setText("Your guess is LOW");
            }
            else
            {
                high = guess - 1;
                resultTV.setText("Your guess is HIGH");
            }
        }
        display(false);
        Toast.makeText(this, "Touch another mobile to continue playing",
                       Toast.LENGTH_SHORT).show();
    }
}
```

As you already know, when the mobiles touch each other, the `createNdefMessage()` method is invoked automatically, since the `setNdefPushMessageCallback()` method is implemented. In this method, five NDEF records are created: previous guesses, possible high value, possible low value, secret number, and the guess status. Finally, a new NDEF message created from these records and returned in order for the message to be beamed to another mobile. The NDEF records are formatted as `TNF_EXTERNAL_TYPE` and the type is formatted as `nfclab.com:guessNumber`. The reason to format the NDEF records as `TNF_EXTERNAL_TYPE` is that you only want your application to handle the incoming beams, since the sent beam is personalized and other applications cannot understand what the incoming data means.

```
@Override
public NdefMessage createNdefMessage(NfcEvent event)
{
    String externalType = "nfclab.com:guessNumber";
    NdefRecord extRecord1 = new NdefRecord(NdefRecord.TNF_EXTERNAL_TYPE,
                                           externalType.getBytes(),
                                           new byte[0],
                                           previous.getBytes());
    NdefRecord extRecord2 = new NdefRecord(NdefRecord.TNF_EXTERNAL_TYPE,
                                           externalType.getBytes(),
                                           new byte[0],
                                           Integer.toString(high).getBytes());
    NdefRecord extRecord3 = new NdefRecord(NdefRecord.TNF_EXTERNAL_TYPE,
                                           externalType.getBytes(),
                                           new byte[0],
                                           Integer.toString(low).getBytes());
    NdefRecord extRecord4 = new NdefRecord(NdefRecord.TNF_EXTERNAL_TYPE,
                                           externalType.getBytes(),
                                           new byte[0],
                                           Integer.toString(target).getBytes());
    NdefRecord extRecord5 = new NdefRecord(NdefRecord.TNF_EXTERNAL_TYPE,
                                           externalType.getBytes(),
                                           new byte[0],
                                new Boolean(found).toString().getBytes());
    NdefMessage message = new NdefMessage( new NdefRecord[] { extRecord1,
                                           extRecord2, extRecord3,
                                           extRecord4, extRecord5}
                                           );
    return message;
}
```

When the data are beamed to the target mobile, you need to implement the onResume(), onNewIntent(), and processIntent() methods. Inside the processIntent() method, you need to process all five NDEF records. Remember that the five NDEF records were previous guesses, possible high value, possible low value, the secret number, and the guess status; and they were beamed in an NDEF message. So, the records are again extracted from the NDEF message and stored in the corresponding variable. Then, the display() method is called, which updates the previous guesses, possible high value, and possible low value on the screen. Since these variables are declared as private, the values inside the class are updated when they are updated in the processIntent() method. The method also checks whether the secret number has been guessed correctly or not by the opponent. If the opponent has found the secret number, the activity displays a message on the screen. Otherwise, the player continues to play the game.

```
@Override
public void onResume()
{
    super.onResume();
    if (NfcAdapter.ACTION_NDEF_DISCOVERED.equals(getIntent().getAction()))
    {
        processIntent(getIntent());
    }
```

```
    }

    @Override
    public void onNewIntent(Intent intent)
    {
        setIntent(intent);
    }

    void processIntent(Intent intent)
    {
        NdefMessage[] messages = getNdefMessages(getIntent());

        for (int i=0; i<messages.length; i++)
        {
            for (int j=0; j<messages[0].getRecords().length; j++)
            {
                NdefRecord record = messages[i].getRecords()[j];
                if(j==0)
                    previous=new String(record.getPayload());
                else if(j == 1)
                    high=Integer.parseInt(new String(record.getPayload()));
                else if(j == 2)
                    low=Integer.parseInt(new String(record.getPayload()));
                else if(j == 3)
                    target=Integer.parseInt(new String(record.getPayload()));
                else if(j == 4)
                    found=Boolean.parseBoolean(new String(record.getPayload()));
            }
            display(true);
            if( found == true )
            {
                resultTV.setText("Your opponent already found the secret number!!! "
                                + target);
                guessET.setFocusable(false);
            }
        }
    }
}
```

Inside the manifest file, you need to add the NFC permission in order to use the NFC adapter.
In order to run the application that receives the beam, you need to declare the required intent
filter in the manifest file. Remember that the application uses a TNF_EXTERNAL_TYPE of
nfclab.com:guessNumber, so you need to declare the related intent filter inside the activity tag.
You can see the manifest file in Listing 8-3.

> **NOTE** *Please see Chapter 5 for detailed information on* TNF_EXTERNAL_TYPE.

LISTING 8-3: Manifest File of the NFC Guess Number Project (NFCGuessNumber\
AndroidManifest.xml)

```xml
<?xml version="1.0" encoding="utf-8"?>
<manifest xmlns:android="http://schemas.android.com/apk/res/android"
  package="com.nfclab.guessnumber"
  android:versionCode="1"
  android:versionName="1.0" >

    <uses-sdk android:minSdkVersion="15" />
    <uses-permission android:name="android.permission.NFC" />

    <application
      android:icon="@drawable/ic_launcher"
      android:label="@string/app_name" >
        <activity
          android:name=".GuessNumberActivity"
          android:label="@string/app_name">
            <intent-filter>
                <action android:name="android.intent.action.MAIN" />
                <category android:name="android.intent.category.LAUNCHER" />
            </intent-filter>
            <intent-filter>
                <action android:name="android.nfc.action.NDEF_DISCOVERED" />
                <category android:name="android.intent.category.DEFAULT" />
                <data android:scheme="vnd.android.nfc"
                  android:host="ext"
                  android:pathPrefix="/nfclab.com:guessNumber"/>
            </intent-filter>
        </activity>
    </application>
</manifest>
```

NFC PANIC BOMB

NFC Panic Bomb is another game developed to play with NFC technology. Inside the game, there is a bomb that will explode after a predefined time. When the game is launched, the application sets the timing to explode the bomb using a random number generator. The person on whose mobile the bomb explodes loses the game. A player can send the bomb to another mobile by beaming it, but he or she first needs to answer an autogenerated question. Only then can that player beam the bomb to the next player. The game ends when the bomb explodes.

When the application is initially launched, a bomb animation is displayed on the screen. A chronometer also starts to count the seconds measuring the lifetime of the bomb. TextView, EditText, and Button objects are displayed on the screen to present the user with a summation problem and enable them to enter and submit an answer to the problem (see Figure 8-15). If the current player cannot give the correct answer to the question, the application asks another question until a correct answer is given (see Figure 8-16). The chronometer runs until a proper answer is given.

FIGURE 8-15

FIGURE 8-16

When the player gives a correct answer to the question, the chronometer stops and a directive to touch to the next player's mobile is displayed on the screen (see Figure 8-17). At this point, the secret lifetime and elapsed seconds are packaged into an NDEF message by the application in order to prepare the data to be transferred.

When a player has answered the question correctly and touched the next player's mobile, the prepared NDEF message is beamed to the target mobile (see Figure 8-18). When the next player receives the beam, the secret lifetime and seconds elapsed are extracted from the NDEF message by the application. Then the player receiving the data tries to give the correct answer to the question and the game continues in this fashion. When the lifetime of the bomb expires, the bomb explodes (see Figure 8-19). See Figure 8-20 for the activity flow diagram of the application.

FIGURE 8-17

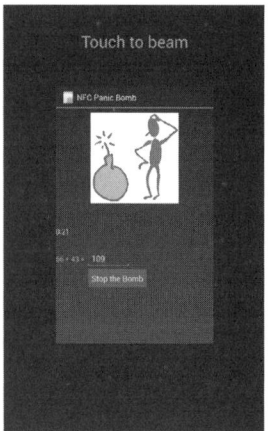

FIGURE 8-18

FIGURE 8-19

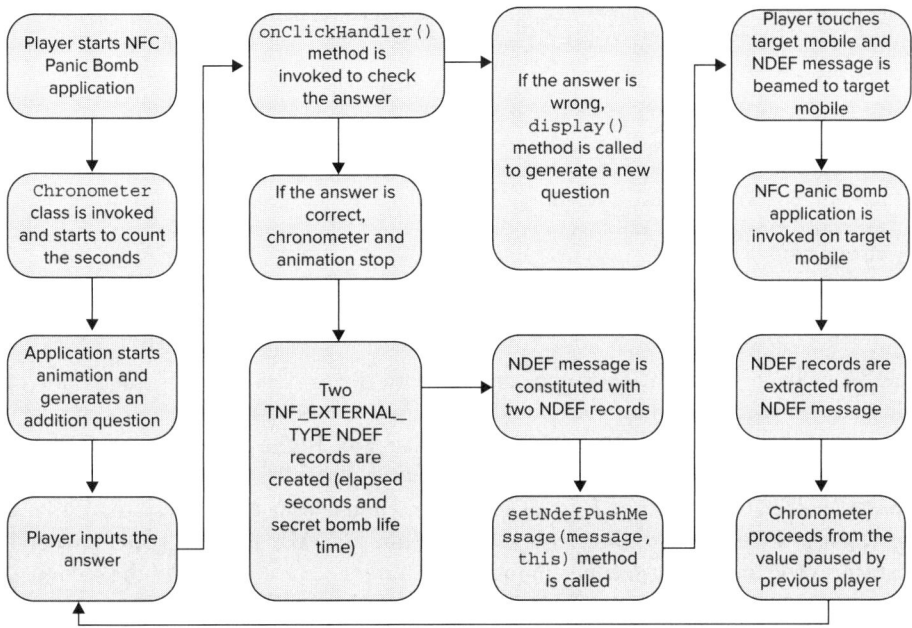

FIGURE 8-20

In the `onCreate()` method, the animation setup is performed, and the display objects are prepared. The `Chronometer` class is used to set the chronometer which stops either when the bomb explodes or when the user gives the correct answer to the question. When the application starts, the chronometer begins at 00:00, but when the bomb is beamed to the next player, that player's chronometer proceeds from the chronometer value of the previous user.

In order to use the chronometer, a `chronometer` object is first started and then a listener is set up in order to check if the total elapsed time reaches the secret lifetime of the bomb. So, you need to check the `elapsedSeconds` variable with the `lifeTime` variable; these store the total spent time and the lifetime of the bomb respectively. If the application is run from scratch, the `elapsedSeconds` variable starts from 0; however, if the application is run via the beam, the variable is updated from the beamed NDEF message. If the elapsed seconds exceed the lifetime of the bomb, you need to explode the bomb. The parameter `elapsedMillis` holds the elapsed time after the chronometer started in milliseconds:

```
chronometer = (Chronometer) findViewById(R.id.chronometer);

chronometer.start();

chronometer.setOnChronometerTickListener(
                                new Chronometer.OnChronometerTickListener()
{
    public void onChronometerTick(Chronometer chronometer)
    {
        chronometer.refreshDrawableState();
        elapsedMillis = SystemClock.elapsedRealtime() - chronometer.getBase();
```

```
            elapsedSeconds = (int) TimeUnit.MILLISECONDS.toSeconds(elapsedMillis);

            if ( ( elapsedSeconds ) >= lifeTime )
            {
                bombIV.setBackgroundResource(0);
                Bitmap bomb = BitmapFactory.decodeResource(context.getResources(),
                                                        R.drawable.after);
                bombIV.setImageBitmap(bomb);
                messageText.setText("Boom!");
                chronometer.stop();
            }
        }
    });
```

The peer-to-peer mode of this application is implemented with the setNdefPushMessage() method.
Remember that when using this method, you should also supply the NDEF message as a parameter.
So you need to use this method when your NDEF message is ready. In the application, you should
call this method when the player correctly answers the question. When the "Stop the Bomb" button
is clicked, the onClickHandler() method is invoked. The method checks the answer and if the
answer is correct, it first stops the chronometer and animation. Then the activity creates two NDEF
records: the first stores the total spent seconds in the elapsedSeconds variable, and the second
stores the lifetime of the bomb in the lifeTime variable. When these NDEF records are created and
the NDEF message is constituted into an NdefMessage object, setNdefPushMessage(message,
this) is called. If the answer is wrong, the display() method is called again to update the
question:

```
    public void onClickHandler(View view)
    {
        if ( view.getId() == R.id.stopButton )
        {
            answer = Integer.parseInt(answerET.getText().toString());
            if ( answer == (number1 + number2) )
            {
                chronometer.stop();
                animation.stop();
                String externalType = "nfclab.com:panicBomb";
                NdefRecord extRecord1 = new NdefRecord(NdefRecord.TNF_EXTERNAL_TYPE,
                                                    externalType.getBytes(),
                                                    new byte[0],
                                    Integer.toString(elapsedSeconds).getBytes());
                NdefRecord extRecord2 = new NdefRecord(NdefRecord.TNF_EXTERNAL_TYPE,
                                                    externalType.getBytes(),
                                                    new byte[0],
                                    Integer.toString(lifeTime).getBytes());
                NdefMessage message = new NdefMessage(new NdefRecord[] {extRecord1,
                                                        extRecord2});
                mNfcAdapter.setNdefPushMessage(message, this);
                Toast.makeText(this,
                            "Touch another mobile to send the bomb",
                            Toast.LENGTH_SHORT).show();
            }
            else
            }
```

```
            displayQuestion();
            answerET.setText("");
            Toast.makeText(this,
                          "Wrong answer!!! Please try again",
                          Toast.LENGTH_SHORT).show();
        }
    }
}
```

When the data are beamed to the target mobile, the `onResume()`, `onNewIntent()`, and `processIntent()` methods handle the received beam:

```
@Override
public void onResume()
{
    super.onResume();
    if (NfcAdapter.ACTION_NDEF_DISCOVERED.equals(getIntent().getAction()))
    {
        processIntent(getIntent());
    }
}

@Override
public void onNewIntent(Intent intent)
{
    setIntent(intent);
}

void processIntent(Intent intent)
{
. . .

}
```

Inside the `processIntent()` method, you need to process the NDEF message that is constituted from two NDEF records: the total spent seconds and the lifetime of the bomb. So, both records are extracted from the NDEF message one by one, and stored into corresponding integer values, which are `elapsedSeconds` and `lifeTime`:

```
NdefMessage[] messages = getNdefMessages(getIntent());
for (int i=0; i<messages.length; i++)
{
    for (int j=0; j<messages[0].getRecords().length; j++)
    {
        NdefRecord record = messages[i].getRecords()[j];
        if( j == 0 )
        {
            elapsedSeconds = Integer.parseInt(new String(record.getPayload()));
        }
        else if( j == 1 )
        {
            lifetime = Integer.parseInt(new String(record.getPayload()));
        {
    }
}
```

These integers are declared as private class attributes so that when you change the value of these variables in the `processIntent()` method, their values in the class are also updated. Furthermore, you need to update the `elapsedMillis` variable by converting from the `elapsedSeconds` variable, and update the chronometer base, in order to start the chronometer from where it stops. You do not need to do anything else in this method, since the other screen elements (chronometer, animation, and so on) are handled in the `onCreate()` method.

```
elapsedMillis = (long) elapsedSeconds * 1000;
chronometer.setBase(SystemClock.elapsedRealtime() - elapsedMillis);
```

Inside the manifest file, you again need to add the NFC permission. `TNF_EXTERNAL_TYPE` is used for NDEF records in this game, with the prefix of `nfclab.com:panicBomb`. So you should also declare the corresponding intent filter in the activity tag. You can see the complete manifest file in Listing 8-4.

LISTING 8-4: Manifest File of the NFC Panic Bomb Project (NFCPanicBomb\AndroidManifest.xml)

```xml
<?xml version="1.0" encoding="utf-8"?>
<manifest xmlns:android="http://schemas.android.com/apk/res/android"
 package="com.nfclab.panicbomb"
 android:versionCode="1"
 android:versionName="1.0" >

    <uses-sdk android:minSdkVersion="15" />
    <uses-permission android:name="android.permission.NFC" />

    <application
     android:icon="@drawable/ic_launcher"
     android:label="@string/app_name" >

        <activity
         android:name=".PanicBombActivity"
         android:label="@string/app_name" >
            <intent-filter>
                <action android:name="android.intent.action.MAIN" />
                <category android:name="android.intent.category.LAUNCHER" />
            </intent-filter>
            <intent-filter>
                <action android:name="android.nfc.action.NDEF_DISCOVERED" />
                <category android:name="android.intent.category.DEFAULT" />
                <data android:scheme="vnd.android.nfc"
                 android:host="ext"
                 android:pathPrefix="/nfclab.com:panicBomb"/>
            </intent-filter>
        </activity>
    </application>
</manifest>
```

SUMMARY

In this chapter, you have seen three use cases in peer-to-peer mode. The first use case is an NFC chatting application, which enables one-to-one online messaging via the peer-to-peer mode of NFC. The application is implemented with `NdefMessageCallback`, and `TNF_MIME_MEDIA` is used for NDEF records.

The second use case is a game in which players try to guess a secret number, which is randomly generated by the application. When a player makes an incorrect guess, that player touches their mobile device to the next player's mobile, and the required values are beamed to the target mobile. The game ends when one of the players finds the secret number. The application is implemented with `NdefMessageCallback`, and `TNF_EXTERNAL_TYPE` is used for NDEF records in this game, with the prefix of `nfclab.com:guessNumber`.

The last use case is a game in which a ticking time bomb explodes after a randomly defined time. Players try to answer an arithmetical question to pause the chronometer of the ticking bomb and beam the bomb to the next player. The secret lifetime of the bomb and the elapsed seconds are packaged into an NDEF message and sent to the target mobile when beamed. The application is implemented with the `setNdefPushMessage()` method, and `TNF_EXTERNAL_TYPE` is used for NDEF records with the prefix of `nfclab.com:panicBomb`. The game ends when the bomb explodes on one of the players' mobiles.

NFC Programming: Card Emulation Mode

After studying the reader/writer and peer-to-peer modes in previous chapters, this chapter discusses NFC technology's card emulation mode. Each NFC operating mode technically includes various properties that need to be studied properly in order to develop applications. Developers must study a different methodology and consider a different set of APIs for each operating mode. Similarly, the ecosystem of possible use cases differs for each mode. When an actor — whether it is a mobile network operator (MNO), financial institute, or IT company — aims to develop a project that requires the use of card emulation mode programming, one or more actors probably need to be involved in the ecosystem of the targeted project.

In contrast to the technical complexity of the scenario of card emulation mode applications, there exist no well-established standards for this mode to enable developers to produce compatible applications. There are some good reasons for the lack of such standards: the complexity of the ecosystem forced application developers to look for alternatives, and the existence of these alternatives confused the developers and became a barrier to a clear solution. These reasons eventually prevented us from exhibiting well-established results.

DEFINITION OF CARD EMULATION MODE

Please remember that the two earlier modes presented different ways for NFC applications to be developed. The reader/writer mode provided ways to communicate with an NFC tag in both directions. It enabled data to be written to the tag, and data to be read from a previously loaded tag. Consequently, projects using reader/writer mode integrated this type of NFC communication. Peer-to-peer mode, on the other hand, enabled communication between two mobile phones in order to exchange data. NFC card emulation mode enables communication between an NFC-enabled mobile phone and a contactless card reader that provides applications based on ISO/IEC 14443 Type A, Type B, and FeliCa communication interfaces. Applications in this NFC mode, therefore, focus mostly on this type of NFC communication. Technically, the idea seems simple — at least as simple as the previous two cases. Unfortunately, in terms of the ecosystem, it is not so simple. There are several reasons for this complication. Card emulation mode applications tend to be directly involved in financial projects, such as credit card, prepaid card, and loyalty card applications. This implies that when an application is to be developed by, say, an IT company, it is essential for the provider of a payment card or loyalty card company, at least, to be involved. When a project involves more than one company, it means that the project is already complex.

> **NOTE** *For detailed information on the NFC business ecosystem, refer to* Near Field Communication (NFC): From Theory to Practice *by Vedat Coskun, Kerem Ok and Busra Ozdenizci (Wiley, 2012).*

BUSINESS ECOSYSTEM

The business ecosystem is where related actors exist in a model in such a way that all the actors communicate in harmony. We leave further explanation and details of the business ecosystem to books like *Near Field Communication* (see the note above), and continue with its most prominent properties as they relate to card emulation mode. In terms of card emulation mode, in particular, the NFC industry can obviously be considered as a new emerging business. The main actors involved in this NFC application mode include MNOs, banking and payment companies, semiconductor and electronic appliance companies including mobile handset makers, software developers, and other merchants, including transport operators and retailers. Because they produce NFC devices, hardware manufacturers affect the card emulation mode; they decide where to implement the secure element (SE), and the way the SE is handled immediately affects the way that card emulation mode applications can be developed. When the Universal Integrated Circuit Card (UICC) is chosen as the place for the SE, MNO companies are immediately central to all potential card emulation applications, because no application can be installed to the UICC without the approval of the MNO that provided the card to the user. For this reason, financial organizations such as banks do not like to see mobile phones with UICC SE but prefer external cards, such as secure digital (SD) cards. The reason for this is obvious; when they aim to develop applications using card emulation mode and install them to the external SEs, it means they do not need to negotiate with the MNO. There is yet another option for storing the SE, which is to integrate it into the mobile phone's embedded hardware. In this case, the manufacturer of the mobile phone may even force all other players out of the game by integrating blocking mechanisms to prevent other actors from installing applications.

STAKEHOLDERS IN AN NFC ECOSYSTEM

NFC card emulation mode potentially includes a wide range of actors. Various NFC ecosystem views are defined by standardization organizations, which differ in terms of defining authority and responsibilities, not to mention benefits. For example, NFC Forum identifies an NFC ecosystem depending on and relating with its member portfolio. On the other hand, the GSM Association (GSMA) describes and visualizes the ecosystem from the subjective point of view of MNOs. However, the key players in an NFC ecosystem are mostly the same in both models (see Figure 9-1).

FIGURE 9-1

More formally, the actors in the card emulation mode ecosystem can be listed as NFC chip set manufacturers and suppliers, SE manufacturers and suppliers, mobile handset manufacturers and suppliers, reader manufacturers and suppliers, MNOs, trusted service managers (TSMs), service providers, merchants/retailers, and customers.

BUSINESS MODELS

There are several business models that can be applied to card emulation mode applications. It is a substantial job to work out the ecosystem details, which include which actors will play a role in the system, what their responsibilities will be, and, of course, their profits. Decisions about the actors, their responsibilities, and their profits are not simple; these are the most challenging issues for projects of this kind. According to Mobey Forum, a number of requirements (i.e., initial, operational, usability, and external requirements) need to be considered while forming business models. When deciding on the business model, questions need to be answered such as:

➤ Who will issue and own the keys to the SE?

➤ Who will manage the life cycle of the SE platform?

➤ Whose over-the-air (OTA) platform will be used for the management of the SE platform?

In relation to these questions, in this section we identify three main issues that determine the business model alternatives for NFC: the SE issuer, the platform manager, and the OTA provider. These issues can also be referred to as functional roles and responsibilities that need to be handled by a single entity or multiple entities in the NFC business model.

Business Model Alternatives

Based on the localization and control of the SE as defined above, three model options emerge: the MNO-centric, the distributed, and the TSM-centric business models.

MNO-centric Business Model

In this model, the MNO issues and controls the SEs. Thus, the MNO performs all the capabilities of the TSM: it owns and manages the loading, installation, and personalization processes, as well as security domain creation on the SE. Service providers have to share their personalization data with the MNO (see Figure 9-2).

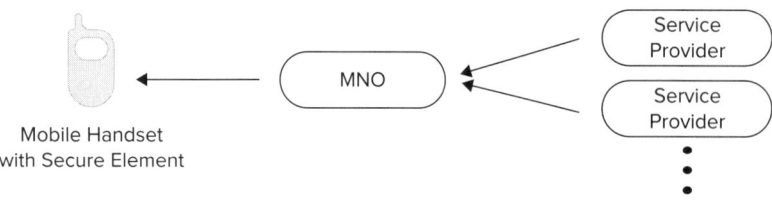

FIGURE 9-2

Distributed Business Model

In some cases, each actor focuses on their specialization area: the MNO is responsible for the distribution and control of the SEs that will be used for hosting the applications; financial institutions are responsible for handling financial issues; and the IT company is responsible for handling the development and maintenance of the application as well as hosting the framework for running the project (see Figure 9-3).

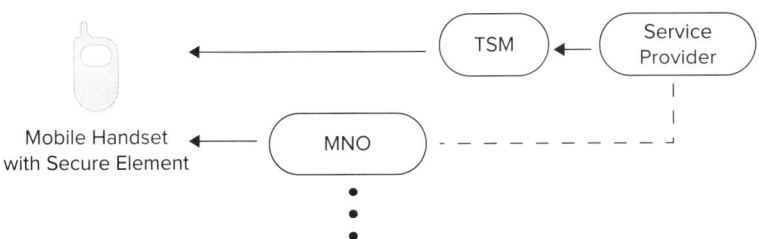

FIGURE 9-3

Today, in most card emulation programming use cases, the distributed business model is preferred. This model actually creates a win-win situation in the ecosystem. However, it has some limitations on the customer side, depending on the SE option. In the case of UICC-based SEs, NFC service can only be offered to a limited number of the service providers' customers, who need to be subscribers of that MNO. To reach more customers, the service provider needs to sign agreements with other MNOs in the market, which creates complexities.

TSM-centric Business Model

For an NFC service, a single TSM-centric business model is actually less complex and thought of as the best model (see Figure 9-4). The involvement of a trusted third entity as the TSM reduces the complexity of the environment, when compared with other models. It also provides fair revenue distribution among MNOs and service providers. In the meantime, a wide range of users benefit from the NFC services, which are provided by different service providers.

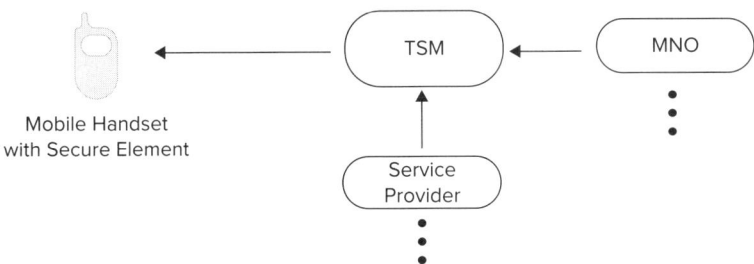

FIGURE 9-4

The number of TSMs may increase depending on the available services and the agreements of the actors in the NFC ecosystem. For instance, NFC-enabled payment and transportation services may use the same TSM platform as well as different TSM platforms. MNOs and service providers wishing to participate in an NFC ecosystem need to sign agreements with the TSM. All entities should share required data with the authorized TSM in the ecosystem. The TSM performs the platform manager's role entirely on behalf of the service providers, by realizing the loading, installation, and personalization processes via its own OTA platform.

Summary of the Business Model Alternatives

Figure 9-5 summarizes all the possible combinations for the three business models discussed above. In all three defined business models, all of the possible SE options can be issued and managed. In the case of the MNO-centric business model, all key indicators are handled by the MNO. In the case of the distributed business model, platform management is performed by the MNO and the service provider. During platform management, the OTA platform that is used may differ depending on the business agreement between the service provider and the MNO. In the TSM-centric business model, the TSM has full authorization over platform management and OTA provision in all phases. Here, the party that issues the SE and the SE option that is used are not important. Authorization of SE management is given over to the central TSM.

SE Option	SE Issued By	Platform Management				Business Model
		Application Loading and Installation		Data Preparation and Personalization		
		Installation	OTA Provider	Personalization	OTA Provider	
All*	MNO	MNO	MNO	MNO	MNO	MNO Centric
All*	MNO	MNO	MNO	Service Provider	MNO	Distributed
All*	MNO	MNO	MNO	Service Provider	TSM	Distributed
All*	MNO	Service Provider	TSM	Service Provider	TSM	TSM Centric
SMC, Embedded Hardware	Service Provider or Retailer	Service Provider	TSM	Service Provider	TSM	TSM Centric

*"All" refers to embedded hardware, SMC, and UICC-based SEs

FIGURE 9-5

One successful story comes from a major MNO in Turkey, called Turkcell. The company equipped many customers with NFC-capable mobile phones and provided NFC-based Cep-T Cüzdan mobile wallets. Cep-T Cüzdan turned customers' mobile phones into mobile wallets to hold credit cards, debit cards, ID cards, tickets, and so on. In addition to allowing the customer to use the mobile phone in contactless terminals, the mobile wallet also offered many services such as coupons, and the ability to transfer funds to other consumers and purchase things online. From an m-commerce (mobile commerce) ecosystem point of view, Turkcell was also successful in integrating different banks in its mobile wallet service. They were the first MNO in the world to provide support for more than one bank in a mobile wallet application.

Another good example of an m-commerce ecosystem is EnStream Company in Canada. EnStream aims to provide solutions for service providers in Canada and to become a common mobile commerce interface between Canada's mobile carriers and the Canadian ecosystem for NFC transactions. They provide an SE management application which allows secure provisioning to SIM- or UICC-based SEs. It provides safe information exchange between Canadian mobile marketplace (carriers) and all credential issuers (e.g., banks, transit commissions, and loyalty companies).

> **NOTE** *For more information about the SE management application of EnStream, please see* `http://www.enstream.com/service_provider_solutions.php`.

Furthermore, another new mobile wallet implementation is provided by Isis. Isis Mobile Wallet was launched in 2012 in Austin, Texas, and Salt Lake City, Utah. Users need to obtain an Isis Ready smart phone which has an NFC chip built in and requires a SIM-based SE card to operate the Isis Mobile Wallet. Users who have an Isis Ready smartphone can use Isis Mobile Wallet easily; first they need to download the application and then load their credit cards (such as American Express, Chase, etc.) and loyalty cards. It also allows PIN protection and remote wallet locking. Isis Mobile Wallet can be used to pay at any Isis Ready merchants who are participating in this trial. Isis aims to expand their mobile wallet application to other regions as well.

According to Mobey Forum, it seems unlikely that MNOs will allow a third party to take over the platform management role of the UICC-based SEs in the case of SIM-based solutions. In fact, for the foreseeable future, it is certain that MNOs will fully or partially take on the role of platform management (i.e., the management of the life cycle of the SE) in business models where UICC-based SEs are used, and MNOs will be the only entities who issue them. Since SD-based SEs are more independent of MNOs and can be applied in all business models, it is possible that they will be preferred by service providers and used more by them. However, from the point of view of the users, deploying the SD cards to the users and increasing the usage of SD cards seems very hard.

In some scenarios, such as ticketing and couponing, which involve additional service providers other than banks and financial institutions, the ecosystem gets even more complex. In an optimal ecosystem, each service provider and MNO needs to sign a business agreement with one or more centralized platform managers in order to have an application uploaded onto SEs. This allows consumers to have access to all available services, switch on and off any service that they choose, switch to another operator at any time, and easily reach a specific service.

General Revenue/Expenditure Flow Model

In terms of the revenue/expenditure aspect of the business model, Mobey Forum has performed an overall analysis of the revenue model. In order to extend this perspective, a useful understanding of simple cash flow among NFC stakeholders is summarized in Figure 9-6.

Stakeholder	Revenue	Expenditure
Customer	➤ Gaining coupons and other benefits from financial and loyalty services	➤ NFC-enabled Mobile Phone ➤ Removable Secure Element; UICC, SMC ➤ Monthly Subscription Fees to Service Providers, and Other Bills
Service Providers	➤ Monthly Fees Paid by Customers as well as Merchants and Retailers	➤ Application Development and Other Application-related Backend Services ➤ Maintenance Services for Applications ➤ Customer Care Services
Mobile Network Operators	➤ Monthly Fees Paid by Its Subscribers	➤ New UICC Issuances ➤ Mobile Network Services ➤ OTA Management Services Depending on Business Model ➤ Billing Subscribers ➤ Customer Care Services
Merchants/ Retailers	➤ Increased Sales (due to new customers or cutomer retention)	➤ Customer Care Services ➤ Bank Fees and Other
Trusted Service Managers	➤ Fees Paid by MNOs and Service Providers Depending on Business Model	➤ TSM Infrastructure Services ➤ OTA Management Solutions Depending on Business Model ➤ Customer Care Services

FIGURE 9-6

CARD EMULATION MODE USE CASE ALTERNATIVES

NFC Card emulation mode enables the implementation of a wide range of applications that potentially require the involvement of many actors. We will now list the most popular potential applications, classified according to their characteristics.

Cashless Payment

It is true that payment systems are a large part of banks' operations, and also help them to enhance their yearly balances. Obviously, most credit and payment card users are also mobile phone users, so why not integrate the payment procedures into mobile phones? In this way, NFC mobile users don't have to carry credit cards as well as their mobiles. Most credit card users carry many credit cards at the same time, so this integration is physically practical as well.

Cashless payment models provide a very practical scenario. When a payment is to be performed, the person with a credit or debit card integrated in their mobile simply touches it and enters a passcode to make the transaction.

Mobile Wallet

If integrating a credit card in a mobile phone is simple, it is equally simple to integrate cash in the same mobile. Be aware, however, that most bank cash storage and transactions are performed digitally or virtually, and not many transactions are performed physically. That way, banks can allow the deposit or withdrawal of any amount of money, because there is almost no risk involved in money processing when it is performed virtually. (This excludes the security mechanisms required to handle and transfer virtual money, of course.)

Given that banks now mostly process cash virtually, why not do the same with regular users who own NFC-enabled mobile phones? This is equally practical. Users can store any amount of money on mobile phones, if the required digital security mechanisms are performed, using NFC mobile phones' card emulation mode. Money can be transferred to the mobile wallet using existing environments, such as bank web portals. The portals need to be modified to enable this functionality, of course. Users can withdraw money from an ATM machine as well, when the machine is equipped with an NFC reader. Instead of physically withdrawing money, you can just touch the mobile to the ATM's NFC reader. When you want to transfer money to a friend, you can do so by using a peer-to-peer mode application that enables this functionality. Stores that only accept cash payments will need to facilitate an NFC reader for this purpose as well.

Ticketing

There have already been many improvements to the process of issuing tickets that have made physical tickets unnecessary. Tickets for flights can be received by mobile via e-mail, and a scanner to read the barcode can monitor its screen when the ticket is displayed on the mobile. NFC is yet another technology that contributes to making mobile ticketing more seamless. After the ticket is imported to the mobile, the mobile phone's NFC capability can be used to transfer the ticket information to the NFC reader that validates the ticket.

Loyalty Cards

Currently, loyalty and membership smart cards have a positive impact on the repeat purchase behavior of loyal customers by stimulating product or service usage. Customers receive cards from different companies, allowing them to get benefits when they make transactions. These might include free miles, points, or coupons, which can be defined as structured marketing efforts.

NFC technology enables people to integrate their daily usage needs into their mobiles securely and eliminates the need for physical loyalty cards to be carried by customers. Customers can simply use loyalty applications on their NFC mobiles and benefit from companies' membership opportunities, earn valuable offers, and so on.

Coupons

Giving discount tickets to potential customers is a very attractive way for companies to persuade them to buy. Some coupons are publicly available via newspapers or leaflets, while others are only acquired, for example, after a purchase has been made. It is important to be able to validate the coupons, and NFC technology makes this easy; the user simply touches their mobile to the NFC reader to validate the coupon.

CARD EMULATION MODE PROGRAMMING

Card emulation programming consists of three different pieces of programming: programming the application in SE, programming the application in Android, and programming the NFC reader.

The SE program is at the center of two devices: NFC reader and Android mobile. The NFC reader communicates with the SE to perform a transaction (for example, a credit card transaction) via the NFC interface. The Android application communicates with the SE to give the user a graphical interface.

In order to program the SE, you need to write an applet and then install it to the SE. The SE is simply a Java Card, so you need to follow the Java Card instructions to write Java Card applets for the SEs. If you already have experience in smart card programming (in other words, Java Cards), you won't have any difficulty programming SEs. For those with no previous experience of Java Card programming, we recommend you visit the official Java Card page and get a book on Java Card, to get started.

> **NOTE** *Please see* `http://www.oracle.com/technetwork/java/javacard/` *for the official Java Card page.*

An open source project named "Secure Element Evaluation Kit for the Android Platform" (SEEK) has been developed in order to program the Android application that communicates with the SE. The project aims to enable the deployment of secure applications on SEs. The project has completed its major modules and is functional. The aim is to integrate the project in the newly deployed Android mobile phones, thus making it possible to communicate with the SEs from Android OS.

> **NOTE** *Please see* `http://code.google.com/p/seek-for-android/` *for the Secure Element Evaluation Kit for the Android platform project.*

PROGRAMMING SECURE ELEMENTS

Applications in an SE communicate with the NFC reader and the Android application. Remember that there are various SEs that can be used: embedded hardware, SD cards, and UICCs. Please refer to Chapter 2, "NFC Essentials for Application Developers," for detailed descriptions of the SEs.

In order to write a program for the SE, you need to use Java Card applets. A Java Card applet is an application on a smart card in the form of Java byte code. When you write a Java Card applet and install it to an SE, the Java Card applet interacts with NFC readers via the NFC interface.

Java Card technology provides a secure application environment for devices that have limited memory and processing capabilities, such as smart cards. One or multiple applications can be deployed to a single card. Many Java Card products also rely on GlobalPlatform specifications.

When you write a Java Card applet, you need to have the keys of the Java Card in order to install the applet to the SE. An embedded hardware SE comes with the mobile phone, and the keys of the SE are generally owned by the mobile handset manufacturer. UICCs, on the other hand, are created and sold by MNOs, so the keys of the UICCs are owned by MNOs. In this situation, you have two options: you can either make an agreement with one of the SE owners to install your application to the SE, or you can distribute SD cards to your customers, which is a difficult and expensive process.

In order to communicate with the SEs from an Android application, you need to use application protocol data units (APDUs). There are two types of APDU: command APDUs and response APDUs. Command APDUs are sent by the smart card reader to send commands to the smart cards; response APDUs are the responses sent by the smart cards.

Command APDUs and response APDUs have a standard format that is defined by the ISO/IEC 7816-4 standard.

In a command APDU, the first byte is the application-specific class of instructions, which indicates the type of the command (see Table 9-1). The second byte is the instruction code that specifies the given command. The third and fourth bytes are the instruction parameters for the given command. Then comes the Lc field, which defines the number of bytes that the command data has. For example, if the command has 4 bytes, you need to specify the length of the command in this field. The next field is the command data, in which you give the command to the SE. Finally, the Le field, which is optional, defines the maximum number of bytes of the response it expects from the applet.

TABLE 9-1: Format of a Command APDU

FIELD NAME	LENGTH IN BYTES	DESCRIPTION
CLA	1	Application-specific class of instructions
INS	1	Instruction
P1	1	Personal instruction parameters for the command
P2	1	Personal instruction parameters for the command
Lc	0, 1, or 3	The number of bytes of command data
Command data	Nc	Command data
Le	0, 1, 2, or 3	Maximum number of response bytes expected (optional)

For example, a command APDU of A0 10 05 02 07 6B 65 72 65 6D 31 32 01 can be broken down into fields as shown in Table 9-2.

TABLE 9-2: An Example Command APDU

CLA	INS	P1	P2	LC	COMMAND DATA							LE
A0	10	05	02	07	6B	65	72	65	6D	31	32	01

The first bytes of the response APDU are the response data (see Table 9-3). The applet returns the requested response with a maximum number of bytes of Le defined in the command APDU. The last two bytes are the processing status of the command. For example, if the operation is performed successfully, it returns 90 00.

TABLE 9-3: Format of a Response APDU

FIELD NAME	LENGTH IN BYTES	DESCRIPTION
Response data	Nr (at most Ne)	Response data
SW1-SW2	2	Command status

For example, a response APDU of 03 90 00 can be broken down as shown in Table 9-4.

TABLE 9-4: An Example Response APDU

RESPONSE	SW1	SW2
03	90	00

PROGRAMMING NFC READER

Programming the NFC Reader is another part of the chain. There are different reader providers and each provider may have different software implementations. In order to program the reader, you should follow your reader supplier's instructions. This part of card emulation mode programming is the most vendor-dependent part. Hence, proprietary solutions have to be followed.

PROGRAMMING ANDROID APPLICATIONS

Android applications in card emulation mode communicate with the SE in order to make GUI-based operations. Some use cases display the card transaction history, card status, changing of the default card, inputting of the card password, and so on. In order to perform these operations, the Android operating system should be able to communicate with the SE to send and receive commands.

In order to promote Android applications to communicate with the SEs, an open source project named "Secure Element Evaluation Kit for the Android Platform" (SEEK) has started with Apache License 2.0. The aim of the project is to make Android an important platform for developing and deploying security-based applications. The project is named as the SmartCard API.

The SmartCard API aims to add necessary modules and APIs to the Android platform and allow applications to access and use SEs. The implementation takes the SIMalliance Open Mobile API specification as a reference. See Figure 9-7 for an overview of the SmartCard API modules.

FIGURE 9-7

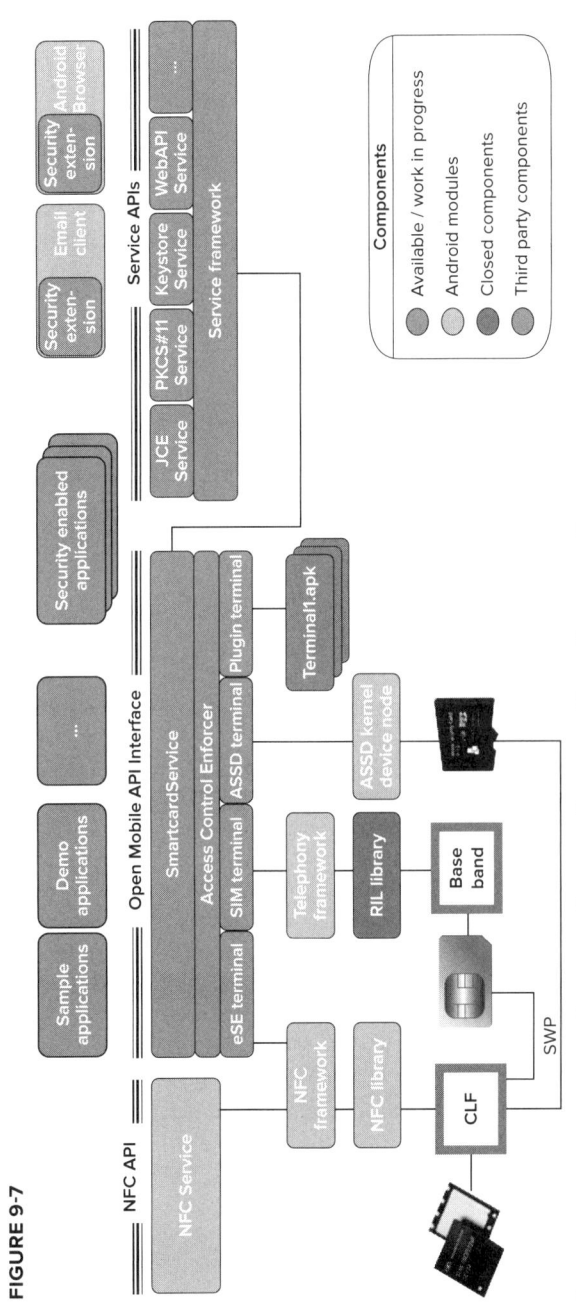

> **NOTE** *See* `http://code.google.com/p/seek-for-android/` *for more information on the SmartCard API.*

Some Android mobile phones have already implemented the SmartCard API to their operating system, but most have not done so yet. Those that already have the SmartCard API are listed at `http://code.google.com/p/seek-for-android/wiki/Devices`.

If you have a mobile device that does not have the SmartCard API, you can rebuild the Android Kernel with the inclusion of the SmartCard API and then flash your mobile phone with the new OS to test the SmartCard API. Note that the mobile phone's warranty may be terminated as a result of this operation.

Enabling Android OS Access to SE

If your mobile doesn't have the SmartCard API, you have two options. The first is to flash your mobile phone with a new Android OS that has an integrated SmartCard API. The second is to use a G&D Mobile Security Card (MSC). This micro SD card enables any Android phone with an SD card slot to use the SmartCard API without flashing the phone.

Flashing the Phone

In order to flash your phone, you need to perform several steps. First of all you need to use Linux or Mac OS. Then you need to set up the requirements in your OS (Linux or Mac OS) in order to build the new mobile Android OS.

> **NOTE** *See* `http://source.android.com/source/initializing.html` *for more information on initializing a build environment.*

In order to build your new Android OS, you need to follow these steps:

1. Get Android sources from Google.
2. Download SmartCard API patch files.
3. Apply the patches.
4. Download the libraries for your mobile phone and extract them.
5. Build the new Android OS for your mobile.
6. Flash your mobile device.

> **NOTE** *See* `http://code.google.com/p/seek-for-android/wiki/` `BuildingTheSystem` *for building a new Android OS with Smart Card API support.*

When you successfully flash your mobile device with the built OS, your mobile is ready to use the SmartCard API.

Note that your mobile phone's warranty may be terminated as a result of these operations.

Using an MSC

A G&D MSC enables any Android mobile phone with an SD card slot to use SmartCard API, without flashing the OS of the mobile phone. When you use this SD card, you will be able to access and test its SE. This is a preferable way to do it for development phones, since you don't need to install a new OS on the mobile phone.

> **NOTE** *See* `http://code.google.com/p/seek-for-android/wiki/` `MscSmartcardService/` *for more information on using an MSC.*

Setting up the Platform

After you have prepared your mobile phone to use the SmartCard API, you need to set up the platform in order to write Android applications that communicate with the SE.

You should have already installed ADT Bundle, which also includes Eclipse IDE, in Chapter 3, "Getting Started with Android." In order to use the SmartCard API, you need to install the Open Mobile API add-on package. To do this, follow these steps:

1. Start Eclipse.
2. Select Window ➪ Android SDK Manager.
3. Click Tools ➪ Manage Add-on Sites.
4. Click the User Defined Sites tab and then click New to add a new repository.
5. Type the following URL into the Add Add-on Site URL dialog box and then click OK:

   `http://seek-for-android.googlecode.com/svn/trunk/repository/addon.xml`

6. Close the Add-On Sites dialog box.
7. Click Packages ➪ Reload to reload the packages.
8. Under Android 4.0.3 (API 15), you will see that Open Mobile API package is added. Select this package and then click Install Package.

Now Eclipse is ready to develop Android applications to access SE.

> **NOTE** *If your mobile phone's API level is lower than 15, you can download and import* `org.simalliance.openmobileapi.jar` *to your project.*

Accessing SE

In order to create an Android application that communicates with SE, follow these steps:

1. Create a new Android project from Eclipse. (See Chapter 4, "Android Software Development Primer," to learn how to create a new Android project.)

2. For the Build Target, select Open Mobile API (Giesecke & Devrient GmbH) (API 15).

3. Select API Level 15 for Minimum SDK.

Now you are ready to create an application to access SE.

Adding Attributes to the Android Manifest File

Inside the Android manifest file, first add a new permission to use the smart card as follows:

```
<uses-permission android:name="android.permission.SMARTCARD"/>
```

Then add a `uses-library` element for Open Mobile API inside the application:

```
<application …>
   <uses-library android:name="org.simalliance.openmobileapi"
       android:required="true" />
</application>
```

Adding Required Codes to the Activity

Import the Open Mobile API in your activity as follows:

```
import org.simalliance.openmobileapi.*;
```

The SE service uses an asynchronous callback mechanism to inform the application when the service is connected. So, implement `SEService.Callback` to the activity and define the callback method `serviceConnected()` as follows:

```
public class MainActivity extends Activity implements SEService.CallBack{
   ...
   @Override
   public void serviceConnected(SEService newSEService) {
   }
```

Then create an `SEService` object within the class:

```
private SEService seService;
```

Inside the `onCreate()` method, use the following code to access the SE:

```
try {
   seService = new SEService(this, this);
}catch (SecurityException se) {
   Log.e("Smart Card Activity Exception",  se.getMessage());
}catch (Exception e) {
   Log.e("Smart Card Activity Exception",  e.getMessage());
}
```

In the onDestroy() method, close the SEService as follows:

```
@Override
protected void onDestroy() {
    if (seService != null && seService.isConnected()) {
        seService.shutdown();
    }
    super.onDestroy();
}
```

From now on, you will start writing code to access the SE. In order to get the available readers, use the following code. If no SE readers are found in the mobile phone, it will terminate the activity.

```
try{
Reader[] readers = seService.getReaders();
if (readers.length < 1)
    return;
```

In order to open a session to an SE, use the following code:

```
Session session = readers[0].openSession();
```

After you open a session to an SE, you need to select an applet from the SE. In order to do this, you need to define the Applet ID (AID). Please refer to Java Card for detailed information on AID. In the following code example, the AID of the applet is 01 02 03 04 05 06 07 08 09 09:

```
Channel channel = session.openLogicalChannel(
new byte[] { (byte) 0x01, 0x02, 0x03, 0x04, 0x05, 0x06, 0x07, 0x08, 0x09,
             0x09});
```

Now, you need to send the command APDU to the SE. Remember that command APDUs are sent to the SE and response APDUs are sent by the SE. In this line of code, you send a command APDU via the transmit() method and save the response to a byte array named responseApdu:

```
byte[] responseApdu = channel.transmit(new byte[] {(byte) 0xA0, 0x10, 0x05,
                                        0x02, 0x07, 0x6B, 0x65, 0x72, 0x65,
                                        0x6D, 0x31, 0x32, 0x01 });
```

Then, you need to close the session to the SE as follows:

```
channel.close();
```

The last thing to do is to process the response APDU. In a response APDU, the last two bytes are the command status, as described earlier. Thereafter, the bytes before the last two bytes are the custom bytes that your applet sent. Inside the command APDU, you have already defined the maximum number of bytes expected from the applet as 1. You may also have defined a different maximum number of expected bytes. In all cases, the last two bytes are the command status. So, you can remove the last two bytes from the response APDU and process the response. Also note that the command status needs to be 90 00, which declares that the execution on the applet is successful. If the command status is different, the execution has not been successful.

You can use the following lines of code to get the command status and the response. After that, you can check that the command status is 90 00 and process the response afterwards.

```
byte[] commandStatus = new byte[2];
System.arraycopy(responseApdu , responseApdu.length - 2, commandStatus,
            0, 2);
byte[] response = new byte[responseApdu.length - 2];
System.arraycopy(responseApdu , 0, response, 0, responseApdu.length - 2);
```

You also need to close the try clause and catch the exception, as follows:

```
}catch (Exception e) {
    Log.e("Smart Card Activity Exception", e.getMessage());
    return;
}
```

You have now accessed an SE from Android in order to send command APDUs and retrieve response APDUs. In order to go further, you need to program the applet with Java Card programming that processes the command APDUs and sends response APDUs. However, the code needed for the Android application to communicate with the SE, sending command APDUs, and retrieving response APDUs, will not change, since this code handles them.

SUMMARY

NFC card emulation mode enables a wide range of important applications that potentially require the involvement of many stakeholders: NFC chip set manufacturers and suppliers, SE manufacturers and suppliers, mobile handset manufacturers and suppliers, reader manufacturers and suppliers, MNOs, trusted service managers (TSMs), service providers, merchants/retailers, and so on.

Currently, some important implementations of NFC card emulation mode are cashless payment, ticketing, loyalty cards, coupons, and turnstile. Due to the complexity of NFC card emulation mode applications in terms of the business ecosystem and technological infrastructure, it is really important to develop suitable business models and processes. Some business models do not encourage the cooperation of all bodies; however, an NFC ecosystem needs various stakeholders and industries in collaboration.

There are three possible collaboration models: the MNO-centric, distributed, and TSM-centric business models. In defining NFC business models, three key issues need to be handled: who will be the SE issuer (depending on the SE alternatives); who will manage the life cycle of the SE and be the platform manager; and, finally, who will be the OTA provider.

Again in terms of technological infrastructure, card emulation Android applications need to be programmed in three parts. The first is to program the application in the SE, the second is to program the Android application, and the third is to program the NFC reader. In order to program the SE, you need to write an applet and then install it to the SE. The SE is simply a Java Card so you need to follow the Java Card instructions to write Java Card applets for SEs.

An open source project named "Secure Element Evaluation Kit for the Android Platform" has been developed in order to program the Android application that communicates with the SE. The project aims to enable the deployment of secure applications on SEs. Some parts of the project have been completed but some are still in progress. When it is completed, the aim is to integrate the project in the newly deployed Android mobile phones.

In order to program the NFC reader, programmers should follow the NFC reader supplier's instructions for that specific NFC reader product and its software. In order to enable SE access from an Android mobile phone using a Smart Card API, there are two options if your mobile does not already have the SmartCard API. The first is to flash your mobile phone with a new Android OS that has an integrated SmartCard API. The second is to use a G&D Mobile Security Card (MSC) that enables an Android phone with an SD card slot to use the SmartCard API without flashing the phone.

URI Prefixes for NDEF

Table A-1 lists the URI prefixes for NDEF records along with their descriptions. These URI prefixes for NDEF records are provided by NFC Forum. The codes are defined in the NFC Forum technical specification titled "URI Record Type Definition."

TABLE A-1: URI Prefixes

DECIMAL	HEX	PROTOCOL
0	0x00	N/A. No prepending is done, and the URI field contains the unabridged URI.
1	0x01	http://www.
2	0x02	https://www.
3	0x03	http://
4	0x04	https://
5	0x05	tel:
6	0x06	mailto:
7	0x07	ftp://anonymous:anonymous@
8	0x08	ftp://ftp.
9	0x09	ftps://
10	0x0A	sftp://
11	0x0B	smb://

continues

TABLE A-1 *(continued)*

DECIMAL	HEX	PROTOCOL
12	0x0C	nfs://
13	0x0D	ftp://
14	0x0E	dav://
15	0x0F	news:
16	0x10	telnet://
17	0x11	imap:
18	0x12	rtsp://
19	0x13	urn:
20	0x14	pop:
21	0x15	sip:
22	0x16	sips:
23	0x17	tftp:
24	0x18	btspp://
25	0x19	btl2cap://
26	0x1A	btgoep://
27	0x1B	tcpobex://
28	0x1C	irdaobex://
29	0x1D	file://
30	0x1E	urn:epc:id:
31	0x1F	urn:epc:tag:
32	0x20	urn:epc:pat:
33	0x21	urn:epc:raw:
34	0x22	urn:epc:
35	0x23	urn:nfc:
36 ... 255	0x24..0xFF	RFU

B

Android NFC Packages

In this appendix, the Android NFC packages are described: the `android.nfc` package in Tables B-1 to B-9 and the `android.nfc.tech` package in Tables B-10 to B-20. Each package's classes and exceptions (if any) are briefly explained with their constructors, fields, constants, and methods. Their API levels are also indicated. Here is a short description of each table:

➤ Table B-1 describes the `android.nfc` package, its classes (`NfcManager` class, `NfcAdapter` class, `NdefMessage` class, `NdefRecord` class, `NfcEvent` class, and `Tag` class), and its exceptions (`FormatException` and `TagLostException` exceptions).

➤ Table B-2 describes methods of the `NfcManager` class of the `android.nfc` package.

➤ Table B-3 describes nested classes (`NfcAdapter.CreateBeamUrisCallback` interface, `NfcAdapter.CreateNdefMessageCallback` interface, and `NfcAdapter.OnNdefPushCompleteCallback` interface), constants, and methods of the `NfcAdapter` class of the `android.nfc` package.

➤ Table B-4 describes fields, constructors, and methods of the `NdefMessage` class of the `android.nfc` package.

➤ Table B-5 describes constructors, constants, fields, and methods of the `NdefRecord` class of the `android.nfc` package.

➤ Table B-6 describes fields of the `NfcEvent` class of the `android.nfc` package.

➤ Table B-7 describes fields and methods of the `Tag` class of the `android.nfc` package.

➤ Table B-8 describes constructors of the `FormatException` exception of the `android.nfc` package.

➤ Table B-9 describes constructors of the `TagLostException` exception of the `android.nfc` package.

➤   Table B-10 describes the `android.nfc.tech` package and its classes (`Ndef` class, `NdefFormatable` class, `IsoDep` class, `MifareClassic` class, `MifareUltralight` class, `NfcA` class, `NfcB` class, `NfcV` class, `NfcF` class, and `NfcBarcode` class).

➤   Table B-11 describes the `TagTechnology` interface of the `android.nfc.tech` package.

➤   Table B-12 describes constants and methods of the `Ndef` class of the `android.nfc.tech` package.

➤   Table B-13 describes methods of the `NdefFormatable` class of the `android.nfc.tech` package.

➤   Table B-14 describes methods of the `IsoDep` class of the `android.nfc.tech` package.

➤   Table B-15 describes constants, fields, and methods of the `MifareClassic` class of the `android.nfc.tech` package.

➤   Table B-16 describes constants and methods of the `MifareUltralight` class of the `android.nfc.tech` package.

➤   Table B-17 describes the methods of the `NfcA` class of the `android.nfc.tech` package.

➤   Table B-18 describes methods of the `NfcB` class of the `android.nfc.tech` package.

➤   Table B-19 describes methods of the `NfcF` class of the `android.nfc.tech` package.

➤   Table B-20 describes methods of the `NfcV` class of the `android.nfc.tech` package.

TABLE B-1: `android.nfc` Package

CLASSES AND EXCEPTIONS	DESCRIPTION	API LEVEL
`android.nfc` package	Provides access to Near Field Communication (NFC) functionality and allows applications to read NFC Data Exchange Format (NDEF) messages in NFC tags	9
`NfcManager` class	High-level manager used to obtain an instance of an `NfcAdapter`	10
`NfcAdapter` class	Represents the local NFC adapter	9
`NdefMessage` class	Represents an NDEF data message that contains one or more `NdefRecords`	9
`NdefRecord` class	Represents a logical (un-chunked) NDEF record	9
`NfcEvent` class	Wraps information associated with any NFC event	14
`Tag` class	Represents an NFC tag that has been discovered	10
`FormatException` exception	Exception	9
`TagLostException` exception	Exception	10

TABLE B-2: NfcManager Class of android.nfc Package

METHODS	DESCRIPTION	API LEVEL
getDefaultAdapter () method	Helper to get the default NFC Adapter	10

TABLE B-3: NfcAdapter Class of android.nfc Package

NESTED CLASSES, CONSTANTS, AND METHODS	DESCRIPTION	API LEVEL
NfcAdapter.CreateBeamUris Callback interface	A callback to be invoked when another NFC device capable of NDEF push (Android Beam) is within range	16
Uri[] createBeamUris (NfcEvent event) method	Called to provide a URI array to push	16
NfcAdapter.Create NdefMessageCallback interface	A callback to be invoked when another NFC device capable of NDEF push (Android Beam) is within range	14
NdefMessage createNdefMessage (NfcEvent event) method	Called to provide an NdefMessage to push	14
NfcAdapter.OnNdefPush CompleteCallback interface	A callback to be invoked when the system successfully delivers your NdefMessage to another device	14
void onNdefPushComplete (NfcEvent event) method	Called on successful NDEF push	14
String ACTION_NDEF_ DISCOVERED constant	Intent to start an activity when a tag with NDEF payload is discovered	10
String ACTION_TAG_DISCOVERED constant	Intent to start an activity when a tag is discovered	9
String ACTION_TECH_ DISCOVERED constant	Intent to start an activity when a tag is discovered and activities are registered for the specific technologies on the tag	10
String EXTRA_ID constant	Optional extra containing a byte array including the ID of the discovered tag for the ACTION_ NDEF_DISCOVERED, ACTION_TECH_DISCOVERED, and ACTION_TAG_DISCOVERED intents	9
String EXTRA_NDEF_MESSAGES constant	Optional extra containing an array of NdefMessage present on the discovered tag for the ACTION_ NDEF_DISCOVERED, ACTION_TECH_DISCOVERED, and ACTION_TAG_DISCOVERED intents	9

continues

TABLE B-3 *(continued)*

NESTED CLASSES, CONSTANTS, AND METHODS	DESCRIPTION	API LEVEL
String EXTRA_TAG constant	Mandatory extra containing the tag that was discovered for the ACTION_NDEF_DISCOVERED, ACTION_TECH_DISCOVERED, and ACTION_TAG_DISCOVERED intents	10
void disableForeground Dispatch (Activity activity) method	Disable foreground dispatch to the given activity	10
void disableForegroundNdef Push (Activity activity) method	This method is deprecated in API level 14. Use setNdefPushMessage.	10
void enableForeground Dispatch (Activity activity, PendingIntent intent, IntentFilter[] filters, String[][] techLists) method	Enable foreground dispatch to the given activity	10
void enableForegroundNdef Push (Activity activity, NdefMessage message) method	This method is deprecated in API level 14. Use setNdefPushMessage.	10
NfcAdapter getDefaultAdapter () method	This method is deprecated in API level 10. Use getDefaultAdapter.	9
NfcAdapter getDefaultAdapter (Context context) method	Helper to get the default NFC adapter	10
boolean isEnabled () method	Return true if this NFC adapter has any features enabled	9
boolean isNdefPushEnabled () method	Return true if the NDEF Push (Android Beam) feature is enabled	16
void setBeamPushUris (Uri[] uris, Activity activity) method	Set one or more URIs to send using Android Beam	16
void setBeamPushUrisCallback (NfcAdapter.CreateBeam UrisCallback callback, Activity activity) method	Set a callback that will dynamically generate one or more URIs to send using Android Beam	16
void setNdefPushMessage (NdefMessage message, Activity activity, Activity ... activities) method	Set a static NdefMessage to send using Android Beam	14

NESTED CLASSES, CONSTANTS, AND METHODS	DESCRIPTION	API LEVEL
void setNdefPushMessage Callback (NfcAdapter .CreateNdefMessageCallback callback, Activity activity, Activity... activities) method	Set a callback that dynamically generates NDEF messages to send using Android Beam	14
void setOnNdefPushComplete Callback (NfcAdapter .OnNdefPushCompleteCallback callback, Activity activity, Activity... activities) method	Set a callback on successful Android Beam	14

TABLE B-4: NdefMessage Class of android.nfc Package

FIELDS, CONSTRUCTORS, AND METHODS	DESCRIPTION	API LEVEL
Creator<NdefMessage> CREATOR field	Creator	9
NdefMessage (byte[] data) constructor	Create an NDEF message from raw bytes	9
NdefMessage (NdefRecord record, NdefRecord... records) constructor	Construct an NDEF message from one or more NDEF records	16
NdefMessage (NdefRecord[] records) constructor	Create an NDEF message from NDEF records	9
NdefRecord[] getRecords () method	Get the NDEF records inside this NDEF message	9
byte[] toByteArray () method	Returns a byte array representation of this entire NDEF message	9

TABLE B-5: NdefRecord Class of android.nfc Package

CONSTRUCTORS, CONSTANTS, FIELDS, AND METHODS	DESCRIPTION	API LEVEL
NdefRecord (short tnf, byte[] type, byte[] id, byte[] payload) constructor	Construct an NDEF record from its component fields	9
NdefRecord (byte[] data) constructor	This constructor was deprecated in API level 16; use NdefMessage(byte[]) constructor	9

continues

TABLE B-5 *(continued)*

CONSTRUCTORS, CONSTANTS, FIELDS, AND METHODS	DESCRIPTION	API LEVEL
`short TNF_ABSOLUTE_URI` constant	Indicates the type field contains a value that follows the absolute-URI BNF construct defined by RFC 3986	9
`short TNF_EMPTY` constant	Indicates no type, ID, or payload is associated with this NDEF record	9
`short TNF_EXTERNAL_TYPE` constant	Indicates the type field contains a value that follows the RTD external name specification	9
`short TNF_MIME_MEDIA` constant	Indicates the type field contains a value that follows the media-type BNF construct defined by RFC 2046	9
`short TNF_UNCHANGED` constant	Indicates the payload is an intermediate or final chunk of a chunked NDEF record	9
`short TNF_UNKNOWN` constant	Indicates the payload type is unknown	9
`short TNF_WELL_KNOWN` constant	Indicates the type field uses the RTD type name format	9
`Creator<NdefRecord> CREATOR` field	Creator	9
`byte[] RTD_ALTERNATIVE_CARRIER` field	RTD alternative carrier type	9
`byte[] RTD_HANDOVER_CARRIER` field	RTD handover carrier type	9
`byte[] RTD_HANDOVER_REQUEST` field	RTD handover request type	9
`byte[] RTD_HANDOVER_SELECT` field	RTD handover select type	9
`byte[] RTD_SMART_POSTER` field	RTD smart poster type	9
`byte[] RTD_TEXT` field	RTD text type	9
`byte[] RTD_URI` field	RTD URI type	9
`NdefRecord createApplicationRecord (String packageName)` method	Create an Android application NDEF record	14
`NdefRecord createExternal (String domain, String type, byte[] data)` method	Create a new NDEF record containing external (application-specific) data	16

CONSTRUCTORS, CONSTANTS, FIELDS, AND METHODS	DESCRIPTION	API LEVEL
`NdefRecord createMime (String mimeType, byte[] mimeData)` method	Create a new NDEF record containing MIME data	16
`NdefRecord createUri (String uriString)` method	Create an NDEF record of well-known type URI	14
`NdefRecord createUri (Uri uri)` method	Create an NDEF record of well-known type URI	14
`byte[] getId ()` method	Returns the variable length ID	9
`byte[] getPayload ()` method	Returns the variable length payload	9
`short getTnf ()` method	Returns the 3-bit TNF	9
`byte[] getType ()` method	Returns the variable length type field	9
`byte[] toByteArray ()` method	This method was deprecated in API level 16; use `toByteArray()`	9
`String toMimeType ()` method	Map this record to a MIME type, or return null if it cannot be mapped	16
`Uri toUri ()` method	Map this record to a URI, or return null if it cannot be mapped	16

TABLE B-6: `NfcEvent` Class of `android.nfc` Package

FIELDS	DESCRIPTION	API LEVEL
`NfcAdapter nfcAdapter` field	The `NfcAdapter` associated with the NFC event	14

TABLE B-7: `Tag` Class of `android.nfc` Package

FIELDS AND METHODS	DESCRIPTION	API LEVEL
`Creator<Tag>` CREATOR field	Creator	10
`byte[] getId ()` method	Get the tag Identifier (if it has one)	10
`String[] getTechList ()` method	Get the technologies available in this tag, as fully qualified class names	10
`String toString ()` method	Human-readable description of the tag, for debugging	10

TABLE B-8: `FormatException` Exception of `android.nfc` Package

CONSTRUCTORS	API LEVEL
`FormatException ()` constructor	9
`FormatException (String message)` constructor	9
`FormatException (String message, Throwable e)` constructor	16

TABLE B-9: `TagLostException` Exception of `android.nfc` Package

CONSTRUCTORS	API LEVEL
`TagLostException ()` constructor	10
`TagLostException (String message)` constructor	10

TABLE B-10: `android.nfc.tech` Package

CLASSES	DESCRIPTION	API LEVEL
`android.nfc.tech` package	Provides access to a tag technology's features, which vary by the type of tag that is scanned	10
`Ndef` class	Provides access to NDEF content and operations on a tag	10
`NdefFormatable` class	Provides access to NDEF format operations on a tag	10
`IsoDep` class	Provides access to ISO-DEP (ISO 14443-4) properties and I/O operations on a tag	10
`MifareClassic` class	Provides access to MIFARE Classic properties and I/O operations on a tag	10
`MifareUltralight` class	Provides access to MIFARE Ultralight properties and I/O operations on a tag	10
`NfcA` class	Provides access to NFC-A (ISO 14443-3A) properties and I/O operations on a tag	10
`NfcB` class	Provides access to NFC-B (ISO 14443-3B) properties and I/O operations on a tag	10
`NfcV` class	Provides access to NFC-V (ISO 15693) properties and I/O operations on a tag	10
`NfcF` class	Provides access to NFC-F (JIS 6319-4) properties and I/O operations on a tag	10
`NfcBarcode` class	Provides access to tags containing just a barcode	17

TABLE B-11: `TagTechnology` Interface of `android.nfc.tech` Package

INTERFACE AND METHODS	DESCRIPTION	API LEVEL
`TagTechnology` interface	`TagTechnology` is an interface to a technology in a tag	10
`void connect ()` method	Enable I/O operations to the tag from this `TagTechnology` object	10
`Tag getTag ()` method	Get the `Tag` object backing this `TagTechnology` object	10
`boolean isConnected ()` method	Helper to indicate if I/O operations should be possible	10

TABLE B-12: `Ndef` Class of `android.nfc.tech` Package

CONSTANTS AND METHODS	DESCRIPTION	API LEVEL
`String MIFARE_CLASSIC` constant	NDEF on MIFARE Classic	10
`String NFC_FORUM_TYPE_1` constant	NFC Forum tag type 1	10
`String NFC_FORUM_TYPE_2` constant	NFC Forum tag type 2	10
`String NFC_FORUM_TYPE_3` constant	NFC Forum tag type 3	10
`String NFC_FORUM_TYPE_4` constant	NFC Forum tag type 4	10
`boolean canMakeReadOnly ()` method	Indicates whether a tag can be made read-only with `makeReadOnly()`	10
`void connect ()` method	Enable I/O operations to the tag from this `TagTechnology` object	10
`Ndef get (Tag tag)` method	Get an instance of NDEF for the given tag	10
`NdefMessage getCachedNdefMessage ()` method	Get the `NdefMessage` that was read from the tag at discovery time	10
`int getMaxSize ()` method	Get the maximum NDEF message size in bytes	10
`NdefMessage getNdefMessage ()` method	Read the current `NdefMessage` on this tag	10
`Tag getTag ()` method	Get the `Tag` object backing this `TagTechnology` object	10
`String getType ()` method	Get the NDEF tag type	10

continues

TABLE B-12 *(continued)*

CONSTANTS AND METHODS	DESCRIPTION	API LEVEL
`boolean isConnected ()` method	Helper to indicate if I/O operations should be possible	10
`boolean isWritable ()` method	Get the NDEF tag type	10
`boolean makeReadOnly ()` method	Make a tag read-only	10
`void writeNdefMessage (NdefMessage msg)` method	Overwrite the `NdefMessage` on this tag	10

TABLE B-13: `NdefFormatable` Class of `android.nfc.tech` Package

METHODS	DESCRIPTION	API LEVEL
`void connect ()` method	Enable I/O operations to the tag from this `TagTechnology` object	10
`void format (NdefMessage firstMessage)` method	Format a tag as NDEF, and write an `NdefMessage`	10
`void formatReadOnly (NdefMessage firstMessage)` method	Format a tag as NDEF, write an `NdefMessage`, and make it read-only	10
`NdefFormatable get (Tag tag)` method	Get an instance of `NdefFormatable` for the given tag	10
`Tag getTag ()` method	Get the `Tag` object backing this `TagTechnology` object	10
`boolean isConnected ()` method	Helper to indicate if I/O operations should be possible	10

TABLE B-14: `IsoDep` Class of `android.nfc.tech` Package

METHODS	DESCRIPTION	API LEVEL
`void connect ()` method	Enable I/O operations to the tag from this `TagTechnology` object	10
`IsoDep get (Tag tag)` method	Get an instance of `IsoDep` for the given tag	10
`byte[] getHiLayerResponse ()` method	Return the higher layer response bytes for `NfcB` tags	10

`byte[] getHistoricalBytes() method`	Return the ISO-DEP historical bytes for `NfcA` tags	10
`int getMaxTransceiveLength() method`	Return the maximum number of bytes that can be sent with `transceive(byte[])`	14
`Tag getTag () method`	Get the `Tag` object backing this `TagTechnology` object	10
`int getTimeout () method`	Get the current timeout for `transceive(byte[])` in milliseconds	10
`boolean isConnected () method`	Helper to indicate if I/O operations should be possible	10
`boolean isExtendedLengthApdu Supported () method`	Standard APDUs have a 1-byte length field, allowing a maximum of 255 payload bytes, which results in a maximum APDU length of 261 bytes	16
`void setTimeout (int timeout) method`	Set the timeout of `transceive(byte[])` in milliseconds	10
`byte[] transceive (byte[] data) method`	Send raw ISO-DEP data to the tag and receive the response	10

TABLE B-15: `MifareClassic` Class of `android.nfc.tech` Package

CONSTANTS, FIELDS, AND METHODS	DESCRIPTION	API LEVEL
`int BLOCK_SIZE constant`	Size of a MIFARE Classic block (in bytes)	10
`int SIZE_1K constant`	Tag contains 16 sectors, each with 4 blocks	10
`int SIZE_2K constant`	Tag contains 32 sectors, each with 4 blocks	10
`int SIZE_4K constant`	Tag contains 40 sectors	10
`int SIZE_MINI constant`	Tag contains 5 sectors, each with 4 blocks	10
`int TYPE_CLASSIC constant`	A MIFARE Classic tag	10
`int TYPE_PLUS constant`	A MIFARE Plus tag	10
`int TYPE_PRO constant`	A MIFARE Pro tag	10
`int TYPE_UNKNOWN constant`	A MIFARE Classic compatible card of unknown type	10
`byte[] KEY_DEFAULT field`	The default factory key	10

continues

TABLE B-15 *(continued)*

CONSTANTS, FIELDS, AND METHODS	DESCRIPTION	API LEVEL
`byte[] KEY_MIFARE_APPLICATION_DIRECTORY` field	The well-known key for tags formatted according to the MIFARE Application Directory (MAD) specification	10
`byte[] KEY_NFC_FORUM` field	The well-known key for tags formatted according to the NDEF on MIFARE Classic specification	10
`boolean authenticateSectorWithKeyA (int sectorIndex, byte[] key)` method	Authenticate a sector with key A	10
`boolean authenticateSectorWithKeyB (int sectorIndex, byte[] key)` method	Authenticate a sector with key B	10
`int blockToSector (int blockIndex)` method	Return the sector that contains a given block	10
`void connect ()` method	Enable I/O operations to the tag from this `TagTechnology` object	10
`void decrement (int blockIndex, int value)` method	Decrement a value block, storing the result in the temporary block on the tag	10
`static MifareClassic get (Tag tag)` method	Get an instance of `MifareClassic` for the given tag	10
`int getBlockCount ()` method	Return the total number of MIFARE Classic blocks	10
`int getBlockCountInSector (int sectorIndex)` method	Return the number of blocks in the given sector	10
`int getMaxTransceiveLength ()` method	Return the maximum number of bytes that can be sent with `transceive(byte[])`	14
`int getSectorCount ()` method	Return the number of MIFARE Classic sectors	10
`int getSize ()` method	Return the size of the tag in bytes: One of SIZE_MINI, SIZE_1K, SIZE_2K, SIZE_4K	10
`Tag getTag ()` method	Get the `Tag` object backing this `TagTechnology` object	10
`int getTimeout ()` method	Get the current `transceive(byte[])` timeout in milliseconds	14

CONSTANTS, FIELDS, AND METHODS	DESCRIPTION	API LEVEL
int getType () method	Return the type of this MIFARE Classic compatible tag	10
void increment (int blockIndex, int value) method	Increment a value block, storing the result in the temporary block on the tag	10
boolean isConnected () method	Helper to indicate if I/O operations should be possible	10
byte[] readBlock (int blockIndex) method	Read 16-byte block	10
void restore (int blockIndex) method	Copy from a value block to the temporary block	10
int sectorToBlock (int sectorIndex) method	Return the first block of a given sector	10
void setTimeout (int timeout) method	Set the transceive(byte[]) timeout in milliseconds	14
byte[] transceive (byte[] data) method	Send raw NfcA data to a tag and receive the response	10
void transfer (int blockIndex) method	Copy from the temporary block to a value block	10
void writeBlock (int blockIndex, byte[] data) method	Write 16-byte block	10

TABLE B-16: MifareUltralight Class of android.nfc.tech Package

CONSTANTS AND METHODS	DESCRIPTION	API LEVEL
int PAGE_SIZE constant	Size of a MIFARE Ultralight page in bytes	10
int TYPE_ULTRALIGHT constant	A MIFARE Ultralight tag	10
int TYPE_ULTRALIGHT_C constant	A MIFARE Ultralight C tag	10
int TYPE_UNKNOWN constant	A MIFARE Ultralight compatible tag of unknown type	10
void connect () method	Enable I/O operations to the tag from this TagTechnology object	10
MifareUltralight get (Tag tag) method	Get an instance of MifareUltralight for the given tag	10

continues

TABLE B-16 *(continued)*

CONSTANTS AND METHODS	DESCRIPTION	API LEVEL
`int getMaxTransceiveLength () method`	Return the maximum number of bytes that can be sent with `transceive(byte[])`	14
`Tag getTag () method`	Get the `Tag` object backing this `TagTechnology` object	10
`int getTimeout () method`	Get the current `transceive(byte[])` timeout in milliseconds	14
`int getType () method`	Return the MIFARE Ultralight type of the tag	10
`boolean isConnected () method`	Helper to indicate if I/O operations should be possible	10
`byte[] readPages (int pageOffset) method`	Read 4 pages (16 bytes)	10
`void setTimeout (int timeout) method`	Set the `transceive(byte[])` timeout in milliseconds	14
`byte[] transceive (byte[] data) method`	Send raw `NfcA` data to a tag and receive the response	10
`void writePage (int pageOffset, byte[] data) method`	Write 1 page (4 bytes)	10

TABLE B-17: `NfcA` Class of `android.nfc.tech` Package

METHODS	DESCRIPTION	API LEVEL
`void connect () method`	Enable I/O operations to the tag from this `TagTechnology` object	10
`NfcA get (Tag tag) method`	Get an instance of `NfcA` for the given tag	10
`byte[] getAtqa () method`	Return the ATQA/SENS_RES bytes from tag discovery	10
`int getMaxTransceiveLength () method`	Return the maximum number of bytes that can be sent with `transceive(byte[])`	14
`short getSak () method`	Return the SAK/SEL_RES bytes from tag discovery	10
`Tag getTag () method`	Get the `Tag` object backing this `TagTechnology` object	10
`int getTimeout () method`	Get the current `transceive(byte[])` timeout in milliseconds	14

`boolean isConnected ()` method	Helper to indicate if I/O operations should be possible	10
`void setTimeout (int timeout)` method	Set the `transceive(byte[])` timeout in milliseconds	14
`byte[] transceive (byte[] data)` method	Send raw NFC-A commands to the tag and receive the response	10

TABLE B-18: `NfcB` Class of `android.nfc.tech` Package

METHODS	DESCRIPTION	API LEVEL
`void connect ()` method	Enable I/O operations to the tag from this `TagTechnology` object	10
`static NfcB get (Tag tag)` method	Get an instance of `NfcB` for the given tag	10
`byte[] getApplicationData ()` method	Return the Application Data bytes from ATQB/SENSB_RES at tag discovery	10
`int getMaxTransceiveLength ()` method	Return the maximum number of bytes that can be sent with `transceive(byte[])`	14
`byte[] getProtocolInfo ()` method	Return the Protocol Info bytes from ATQB/SENSB_RES at tag discovery	10
`Tag getTag ()` method	Get the `Tag` object backing this `TagTechnology` object	10
`boolean isConnected ()` method	Helper to indicate if I/O operations should be possible	10
`byte[] transceive (byte[] data)` method	Send raw NFC-B commands to the tag and receive the response	10

TABLE B-19: `NfcF` Class of `android.nfc.tech` Package

METHODS	DESCRIPTION	API LEVEL
`void connect ()` method	Enable I/O operations to the tag from this `TagTechnology` object	10
`NfcF get (Tag tag)` method	Get an instance of `NfcF` for the given tag	10
`byte[] getManufacturer ()` method	Return the Manufacturer bytes from tag discovery	10

continues

TABLE B-19 *(continued)*

METHODS	DESCRIPTION	API LEVEL
`int getMaxTransceiveLength ()` method	Return the maximum number of bytes that can be sent with `transceive(byte[])`	14
`byte[] getSystemCode ()` method	Return the System Code bytes from tag discovery	10
`Tag getTag ()` method	Get the `Tag` object backing this `TagTechnology` object	10
`int getTimeout ()` method	Get the current `transceive(byte[])` timeout in milliseconds	14
`boolean isConnected ()` method	Helper to indicate if I/O operations should be possible	10
`void setTimeout (int timeout)` method	Set the `transceive(byte[])` timeout in milliseconds	14
`byte[] transceive (byte[] data)` method	Send raw NFC-F commands to the tag and receive the response	10

TABLE B-20: `NfcV` Class of `android.nfc.tech` Package

METHODS	DESCRIPTION	API LEVEL
`void connect ()` method	Enable I/O operations to the tag from this `TagTechnology` object	10
`NfcV get (Tag tag)` method	Get an instance of `NfcV` for the given tag	10
`byte getDsfId ()` method	Return the DSF ID bytes from tag discovery	10
`int getMaxTransceiveLength ()` method	Return the maximum number of bytes that can be sent with `transceive(byte[])`	14
`byte getResponseFlags ()` method	Return the Response Flag bytes from tag discovery	10
`Tag getTag ()` method	Get the `Tag` object backing this `TagTechnology` object	10
`boolean isConnected ()` method	Helper to indicate if I/O operations should be possible	10
`byte[] transceive (byte[] data)` method	Send raw NFC-V commands to the tag and receive the response	10

INDEX